Model Railroad El

CW01654606

Rutger Friberg

Published by
Allt om Hobby

Model Railroad Electronics 4

Author: Rutger Friberg, who has also built the prototypes and made the drawings

Original title: Elektronik för modelljärnvägen

Editing and Layout: Peter Liander, Freddy Stenbom and Mattias Stenbom
Cover Photo: Jens Dahlström
Back Cover Photo: Maria Stenbom
Other photos by the author, Jens Dahlström or from the *Allt om Hobby* archives

The series Model Railroad Electronics, by Rutger Friberg, includes the following titles:

Model Railroad Electronics
1994, ISBN 91-85496-73-1

Model Railroad Electronics 2
1995, ISBN 91-85496-82-0

Model Railroad Electronics 3
1996, ISBN 91-85496-91-X

ISBN 91-85496-51-0

Publisher: Allt om Hobby AB
Box 90 133, S-120 21 Stockholm, Sweden
Tel: +46-8-999 333, Fax: +46-8-99 88 66.
E-mail: freddy.stenbom@hobby.se
Internet: http:/www.hobby.se

Printed in Sweden by Gummesons Tryckeri AB, Falköping

Table of Contents

(see listing on page 4)

Introduction

Writing a book is a great challenge. I have much respect for the writers that can produce those combinations of technical solutions, ideas and experiences, and do so with an element of entertainment.

To write a fourth book on the same subject feels in a way, like an even greater challenge than the earlier efforts. The third book was written five years ago. These have been five years of dramatic technical development in areas such as information technology and electronics. It is a challenge to try to reflect these new technical developments in a book such as this.

To combine the model railroad with electronics projects is a fixation that I just can't let go. The model railroad is the perfect "platform" for my ideas and construction projects. It is a model of reality with just the right amount of exactness and manipulative qualities that can be built at home and at a reasonable price.

That the reader, through the book, will be able to combine hobby activity with the search for knowledge is one of my goals in writing. I especially hope that the younger generation will be attracted by the information technology developments and the exciting union of the personal computer with one's own model railroad where trains are program-controlled. To be able to learn a bit of the basics of electronics by building various model railroad constructions can be the step needed in making a career choice within computer technology or electronics later in life.

Who is this book directed to? The beginners have their own section with equipment and material purchasing advice, soldering instruction followed by extra explanations of the simpler construction projects. The average hobbyists have dozens of new projects to build, with finished circuit board layouts and troubleshooting instructions provided as background support. The pros can try the more complex projects with both analogue and digital technology as well as several alternatives using the new surface mounting technique. For the computer technician, there are train-control programs.

It is important to remember that one need not be an electronics engineer to benefit from the book's projects. There is a mixture of theory and practice that I hope is both comfortable for you and easy to read. Many of the projects can be of use to those not at all interested in model railroading, but rather with completely different hobby interests.

Mixed with the heavier sections covering various themes are several side-bar tips. Often, the more simpler circuits are presented here. At the same time, the reader is encouraged to build circuits of his or her own design, using the book's final, rather detailed chapter covering component descriptions as a reference. If contrary to expectations, a module does not work the first time, have a little patience — relax awhile with something else — and then make a systematic examination of the circuit board. Often, annoying faults such as incorrectly positioned components and excessive solder, are the cause for malfunction. There are plenty of troubleshooting instructions in the book to help you in remedying defects.

All and all, the book is directed to both AC and DC modelers. It has projects for all scales, from Miniclub to G scale.

Finally, my thanks to larger parts of my circle of friends that have contributed with both advice and project ideas, and built test constructions of several of the modules.

This book would never have come about without the support of my dear family — my wife Susanne and the teenagers, Sofia and Oscar (my toughest critic). Thank you!

Rutger Friberg

Rutger Friberg is a frequent lecturer in electronic's and model railroad projects. Here is such an occasion at the Älvsjö Hobby Show (nov 94). He has also presented more than 30 clinics at NMRA conventions in USA.

Information Sources

There are loads of sources for learning more on electrical power and electronics. Among these are electronics magazines, featuring both articles covering general construction projects and electronic theory. Hopefully, you have a well-stocked library in your area—countless books have been published that will help you on your way. If your interests are in combining electronics with model railroading, keep in mind that this book is just one in a series.

ARE THERE COMPONENTS AVAILABLE LOCALLY?

It is often wise to shop at your local hobby or electronics store. There you can get personal and direct help. If you have no local place of purchase, below is a guide to well-known suppliers. The list was made in the spring of 1997, and I hope that it is correct even later on.

FILMS AND CIRCUIT BOARDS

Most of these companies are general suppliersand have a good assortment of components, but there are also a few specialists which are mentioned further on. One example is Tekno-Bygg in Åryd, Sweden. Besides components this company also offers originals on film if you want to make your own circuit boards (copying from the book is possible but it often requires retouching). You can also get circuit boards (drilled or not) from this company or from Hobbycenter in Gothenburg, Sweden.

SPECIAL DIGITAL COMPONENTS

To many of the 3R-digital projects there are special components and DIY kits available at AV-service in Nacka, Sweden (those marked BVE and BERF). The purpose of this is that you from this company can buy special components for 2R-digital (mostly decoder CPUs).

2R DIGITAL COMPONENTS

In this book there are several DIY-projects according to the NMRA DCC protocol. The adjacent table with components suppliers indicate sources that normally carry most of these used components. A few special IC:s are used and produced by Lenz GmbH in Germany (fax= +49-6403-5332). A Model Railroad club in Gothenburg, Sweden has obtained substancial skills in this digital field and also carries stock of these components for sales in smaller quantities. The club Model Railroaders in West (MRW) can be reached through fax (+46-31-775-2910) or via their homepage "www.hobby.se/MRW/MRW.HTM"

If you need in-depth info on the Motorola IC-series 14502X then please consult:
"http://motserv.indirect.com/logic/cmos/overview.html".

Files like "mc145026.pdf" are available there.

SMALL INCANDESCENT LAMPS

At the end of the book you will find a separate table just on different Incandescent Lamps for a variety of applications and voltage levels. Most of the general electronics suppliers stock some of these.

Alternatively you may contact a specialist that carries most of the lamps for Modelrailroad and general hobby use. It is Miniatronics Corp in Deer Park, NY (Fax 516-242-7796).

Electronic Components Suppliers list:

Digi-key Corporation
701 Brooks Ave South
Thief River Falls
MN 56701-0677
USA
tel 800-344-4539

Jamenco
1355 Shoreway Road
Belmont
CA 94002-4100
USA
tel 800-831-4242

Maplin Electronics
P.O.Box 3
Rayleigh
Essex SS6 8LR
England
fax 702-553-935

Ericsson Electronic Services
P.O.Box 719
S-194 27 Uppl-Väsby
Sweden
fax +46-8-757 5101

Tekno Bygg
P.O.Box 3141
S-360 43 Åryd
Sweden
tel +46-470-774-319

Hobbycenter
P.O.Box 4021
S-400 40 Gothenburg
Sweden
fax +46-31-125320
"www.hobbycenter@hobbycenter.se"

AV-service AB
P.O.Box 60
S-131 06 Nacka
Sweden
tel +46-8-716 99 84

MINITIPS

A large number of small tips are spread throughout the book. You will find the following minitips in this book:

Tool Tips for Beginners

This section is for the beginner that hasn't yet obtained suitable tools or is in need of complementing the basic kit. We will only take up the most necessary tools here and omit the generally more specialized help modules, tweezers, magnifying glasses, diverse work-area illumination, etc., that are so important in achieving good results. Let these investments come later, when you are more at home with this sector of the hobby.

There are corresponding, brief tool introductions in each of the previous books of this series. But in consideration of the pace of both technical and economic developments (most things are increasing constantly in price), this latest "update" is included.

Beginners Tools kit

Tool	Scandinavian Sources (Country code = +46)* Hobbex, Sweden Fax 33 140 275	Elfa, Sweden Fax 8 730 30 88 www.elfa.se	F:a Clas Ohlson Fax 247 44 555 order@clasohlson.se
screwdriver	kit 46-44 33	kit 80-402-14	kit 30-3391
pliers	kit 46-36-09	No. 80-462-94	kit 30-4277
stripper	–	No. 80-514-01	No. 20-226
multitest	Analog DC-AC-Res No.56-10-75	Digital DC-AC-Res No. 76-719-02	Analog DC-AC-Ohm No. 32-3939
Solder.stn	30 W st-alone 65-13-87	Weller station 82-300-13**	25W solder pen 30-2105**
Tin remove	No.46-24-73	No. 82-902-07	No. 30-2074
Minidrill	No. 46-51-57	No.80-676-05	No.40-4425***
Drilltable	No. 46-51-40	No.80-676-88	table included

remarks:
* see p. 4 for International Source listing
** these are 220 V versions
*** professional PCB drill

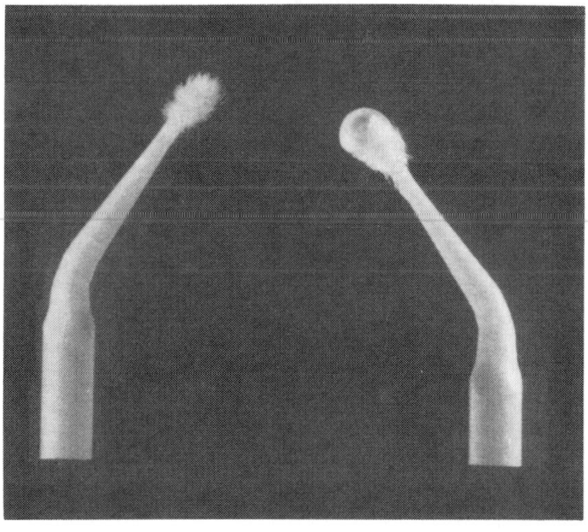

The need of the tools marked in the adjacent table on this page is obvious when doing electronic projects work. Added I personally always have a number of Microbrush bendable applicators at hand for any general assistance. Typical tasks are cleaning connectors and relay-contacts; or removing excess solvent or glue; or applying solder tin paste. etc.

MINITIPS (continued from page 4)

Tips for Electronics Novices

This section is directed to those of you that have just begun to take an interest in electronics and its application potential to the model railroad. If you've already spent time with a soldering iron in your hand, jump to the next section.

As a beginner, one falls into traps of every conceivable description. It is important that you don't get stuck. You must experience the pleasure of success on the road to learning this fascinating hobby—a hobby that is both instructional and entertaining. Here are a few tips that I would like to share with you so that you'll be able to feel that you're on to something worthwhile.

START WITH SIMPLE PROJECTS

A general word of advice is to save the large and complex construction projects until later. Wait with the surface component projects until you have more experience with soldering and don't buy components for all of the projects at once. In step with investment levels, you are unconsciously placing higher demands upon yourself. So start slowly.

In this fourth book are several construction projects that require just a few components. If you are just starting out, many of the projects in the first and third books can be recommended. Note in passing that the second book is specifically directed to three-rail digital technique (3-R digital).

SET UP YOUR OWN "SOLDERING CORNER"

It is practical to have your own little work area when working with electronics. It's often best to work close to the layout (if you have or are constructing one) in that soldering, disassembly and trouble shooting are often done both at the workbench and on the layout.

An important feature of the electronics work area is the floor, or to be more exact, what you have on the floor. All types of carpeting should be avoided because they radiate static electricity (you get a shock now and then). The charges contain little energy, but high voltages. These charges can devastate electronic components, especially prior to soldering them in place.

Linoleum and wood floors are fine.

If you want to be really sure of keeping these transients away from the components, a conducting plastic mat (made by 3M) can be used as a working surface. It is connected to ground. This often means that a lead is drawn to the closest metal pipe or to a metal rod, driven into the ground. Don't be tempted into using the ground in a wall outlet. You have to be a pro to do that.

USE AN APPROVED MODEL RAILROAD TRANSFORMER

Many projects are powered by batteries, either common batteries or small rechargeable batteries. Others projects are designed for powering from some sort of line voltage device. The beginner should keep to approved, factory manufactured equipment. Model railroad transformers are especially suited for powering most of the projects in this book. Many hobby shops offer good used transformers for a fraction of the cost of new ones. Never experiment using line voltages.

As long as we are on the subject of safety, the limit for non-fused loads on a single line in Sweden is 50W. With the usual maximum of 14V-18V used on model railroads, approximately 3A is the maximum current that passes through a lead. This puts us in the 42W-54W range. Divide the layout into separate power blocks if you have a large, combined power consumption. If you use higher effects, there is a risk for over-heating and in the worse case, fire.

ORGANIZE YOUR WIRING

Generally, the best way to avoid overloads and short circuits on the layout is to keep things organized under the baseboards. Put in that extra effort to draw wires straight, use color coding and appropriate attachment fixtures (see the sketch). The risk is otherwise great that you will create a rat's nest, where you will never find what you are looking for. Proper documentation reduces dramatically time spent on trouble shooting. In a wiring notebook, enter wiring layout and connection point information. Discipline yourself to keep the notebook current—enter changes.

The most common, and probably the most boring electrical work is lead connection (and it is often carried out under the layout). This may sound banal, but a primary cause of electrical problems is improper connections. All connections should be soldered. Twisting the leads together is all right for the moment but rather quickly, oxidation decreases conductivity. Conductivity loss brings voltage drops at the connections. This is a difficult fault to detect. Avoid this problem by always soldering connections.

INSULATE WITH SHRINK TUBING OR USE SCREW TERMINALS

Next, the connections should be insulated. I avoid using electrical tape, even if it is better than no insulation at all. One way of making connections is to use screw terminals. These are often mounted in plastic encasing strips of various dimensions that can be clipped to the desired length. This is a good and inexpensive solution.

Myself, I've gone over more and more to shrink tubing. The tubing is sold by the foot (several sizes are available in electronics shops). It can be cut in appropriate lengths (commonly, about a quarter inch) and threaded over the connection point. The tubing is heated (with a hair dryer or other source of warm air) and the tubing quickly shrinks around the enclosed material. Naturally, the tubing is of insulating material.

USE A SEPARATE LEAD FOR GROUND

There is a special case when insulation is not used. Under the baseboards and around the entire layout, install a heavy (about 1/16") brass wire, wrapping it around nails or screws to keep it taut. This lead is the layout's ground

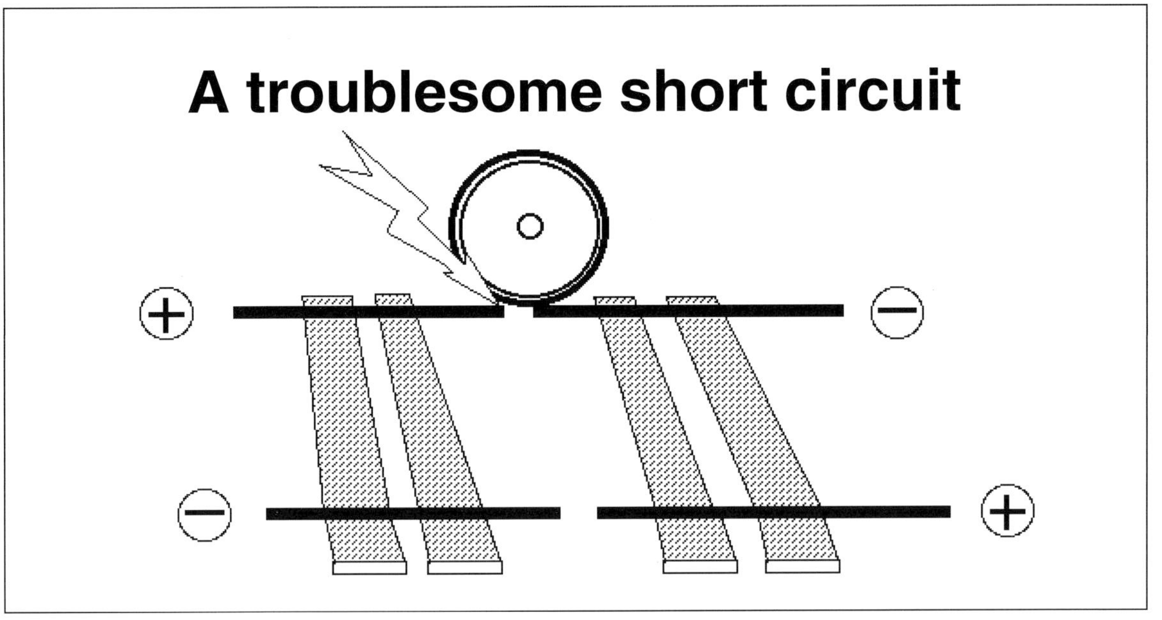

conductor—or neutral on AC systems—and is connected to a central feed. The advantage with this is that wiring is simplified. One pole from nearly all loads can be connected to the nearest point on the ground conductor by soldering directly to the brass wire. Other leads are insulated as usual and with well-defined connection points. This eases trouble shooting enormously.

Finally, I would like to direct your attention to a separate section in the book covering the purchase of suitable equipment and tips on soldering (especially surface components). You will find more on soldering techniques and equipment in the first and second books of this series.

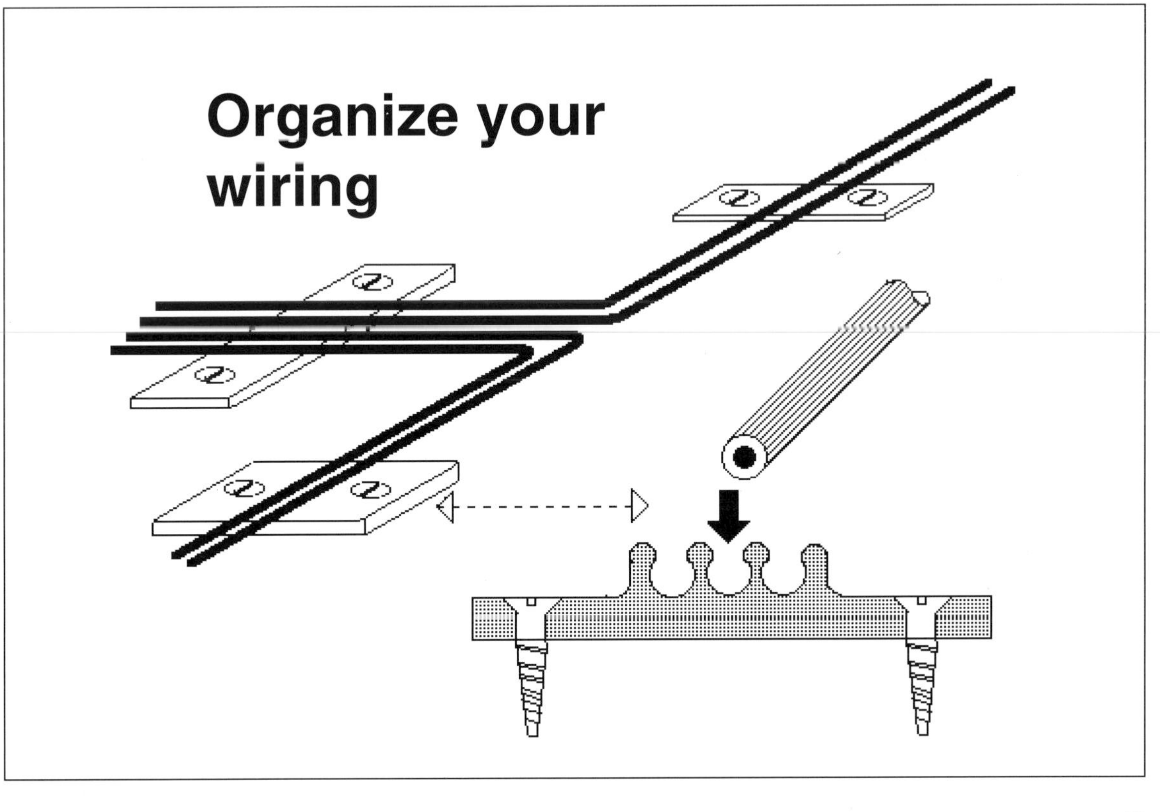

Instrument tips

Instruments are a science unto themselves. There is no upper limit for how much a sophisticated instrument can cost. Because electronics work usually deals with quantities that are impossible to see and feel, it is difficult to estimate values—take voltage, current and resistance as examples. Instruments are definitely a necessity. Contrast this with the more "physical hobbies" where one can measure sizes and weights—quantities that can be estimated.

We must have some sort of quality classification of instruments so that we know how much (or how little) we can rely on the measured results. At the same time, it should also be noted that common resistors have a tolerance of ±10% and if you need 5% or 1% tolerance resistors, you must ask for them specifically. With a good ohmmeter, you can measure your way through the "ten-percenters" until you find one that has the exact value desired—if the instrument is accurate, that is.

On better analogue instruments, the most accurate measurement range is often marked on the instrument itself. This means that precision can be improved by simply switching to an alternate range. A good analogue instrument also incorporates a small mirror under the pointer as an aid in reading the instrument at a right angle to the scale. These features may seem extraneous, but it doesn't cost anything to know about them.

Technology has now advanced to the point that even inexpensive instruments can manage 1-2% accuracy within the common measurement ranges. This degree of accuracy can be difficult to attain when measuring megaohms but here, 3-5% accuracy is acceptable.

Good instruments for hobby use should be in the 1-2% range. The range for professional instruments is 0.01-0.5%.

The most useful instrument for electronics is the multi-purpose instrument. It measures quantities such as volts, amperes and ohms. Basic instruments can only measure DC, but for model railroad applications, AC capability is also needed. This feature is invaluable for working with Märklin's analogue power system and common transformer AC lighting ouputs, to name just a few examples. The ability to measure everything from millivolts and all the way up to 500V is recommended.

The capability of measuring microampere leakage currents as well as motor and battery charging currents of 2-4 amps is also desirable.

Because scales do not provide precise readings, resistances in the 0-1W are not possible to measure. Most instruments manage the upper ranges well.

Although the digital verses analog instrument question can be dismissed as a matter of taste, there are practical aspects to be considered. The multiple scales on an analog instrument can be confusing for the beginner and lead to faulty readings. On the other hand, if a fluctuating value is measured, the value is easier to interpret on an analog display. More expensive digital instruments feature a "hold" function and an adjustment for setting measurement intervals to ease reading values of this type.

The second instrument on my list is a logic probe for measuring values in both TTL and CMOS circuits. It should be able to measure absolute values (high and low levels) as well as pulses—not the total or frequency but just that pulses are present. Slow pulse trains are indicated by a slow LED blink rate. Higher frequencies should make the LED blink faster. There is a project in this book for building a perfectly satisfactory logic probe. If you would rather buy one, a unit matching the specifications in the instrument table can be found for less than $50.00.

If you can afford it, I would recommend buying an oscilloscope. To put it simply, it's like a little TV with measuring cables. The screen provides a graphic display of the signals. This is valuable. Usually, oscilloscopes use a cathode ray tube (as in televisions, except smaller) but oscilloscopes with LCD displays are also available.

In our price class, as shown in the table, we can keep to analog measurement principles. More expensive instruments work digitally and allow us to "replay" sequences, for example. Next time around, I may go into more detail on how to work with and how to gain full advantage of oscilloscopes.

Although function generators may be extravagant, the ability to transmit pulses of various types for signal generation, simulations and troubleshooting constructions is helpful. Input signals are produced and sent to the circuit. The signals can be followed using a logic probe or oscilloscope at various points within the tested circuit. If the signal chain is suddenly broken, you can probably isolate the faulty component or connection point.

At the bottom of the list is a component tester. It is always good to be able to test components. While you can also do this with an oscilloscope, it's nice to be able to avoid the temporary connections inherent with oscilloscope testing methods. With a component tester, components are connected directly to clips on the instrument casing. The tester is especially good for those of you that salvage older electronic devices for parts. It will certainly pay for itself quickly.

Instruments		
Oscilloscope	20 MHz, X and Y-inputs, also ext. trigger	$400
Function generator	Sinus and square wave,20 Hz-100 KHz	$100
Logic probe	20 MHz, TTL and CMOS logic, 1MΩ impedance	$40
Component tester	resistors,capacitors,diodes,batteries, LEDs,transistors,thyristors	$110
Total	in the area of	$650

Quick guide for determining parallel resistance

The total resistance for series connected resistance is easy to calculate—just add the individual resistor values. It takes considerably more effort to calculate parallel resistance. The formula provided shows how to do it.

If an approximate value will suffice, there is a simpler method. We draw a line on the accompanying graph between the R1 and R2 scales from points matching the values of the parallel resistors. The total resistance can be read where this line intersects the Rx scale. In the example, represented by the dashed line, we have parallel connected resistors of 150 kΩ and 100 kΩ. The total resistance is 60 kΩ. Naturally, the corresponding calculation for ohms or megaohms can also be made.

The guide is also applicable for capacitors connected in series. The same basic formula is used as in calculating parallel resistance.

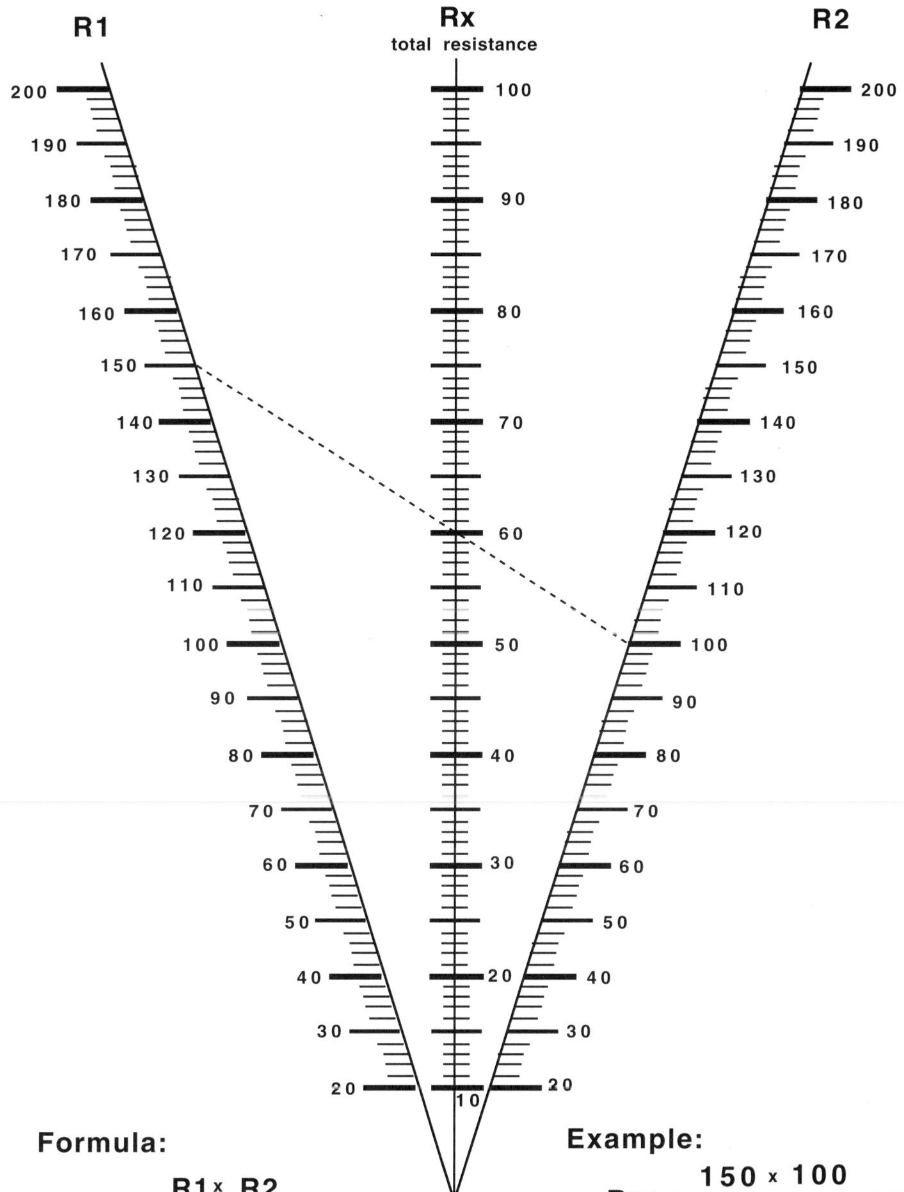

Formula:

$$Rx = \frac{R1 \times R2}{R1 + R2}$$

Example:

$$Rx = \frac{150 \times 100}{150 + 100} = 60$$

Tips for working with surface components

One often comes across the designations SMD (Surface Mounted Device) and SMT (Surface Mount Technology) used as practical abbreviations when referring to surface mounting, its technology and components. I often use these abbreviations and I hope that you will bear with me.

Technology is constantly moving forward. If we brush aside television picture tubes, the tube age feels very remote from today's perspective. Reduction in size is a continual trend in electronics.

Thanks to SMT, we can pack functions into small locomotives and cars that wouldn't even have considered during the '80s. The builder must make a choice. Some of the book's projects can only be executed with SMT if they are to be used in the HO or N scale environments. In certain cases, SMT can even make possible module construction for Z scale.

On the other hand, if you only run trains in O, 1 or G scales, there is plenty of space to build using "traditional" components only.

Surface components are small—*very* small, you might say. This is why we can build so compactly. How does one go about handling and soldering these miniature components? Before we get into that, a few words on surface component storage. I have a plastic case with 130 removable compartments (each with a lid) where I can store and keep my SMD components in order. The component values are written on labels attached to each lid. I make it a habit to never have more than one compartment open at a time. If components are mixed, it is very difficult to sort them out again. And when it comes to capacitors, they usually are not marked at all. The only way to identify an unmarked capacitor is through measurement.

Although single components are available, it often pays to collect your component needs into a single, large volume order. A good part of component costs are in handling. Break points for volume pricing are usually at 10, 25 and 100 items.

Work with SMD can be divided into three phases: physical installment, soldering and later, function testing.

Installation work demands a certain amount of patience and concentration, especially at the beginning. Just let me say now, that experience greatly reduces the degree of difficulty and with simple means, one quickly attains functioning results. The pros often use automatic collection and installation machines. We'll keep to manual labor. There are however, many aids to ease installation. A little

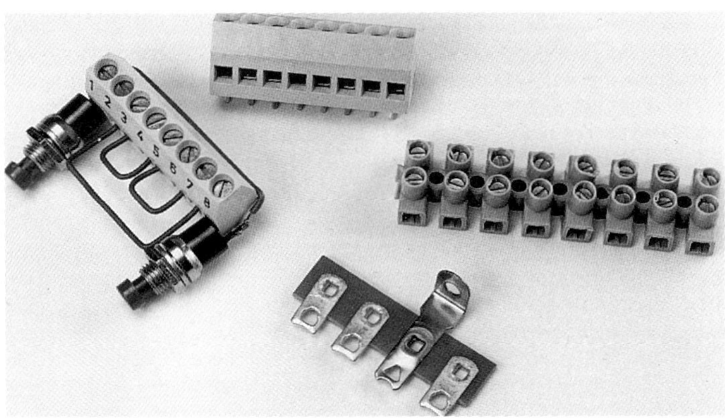

Here are a few of the available terminal strips that can be used for systematic connection of leads. They can be used to splice wires or for parallel power feeds. The strip to the far right is a screw terminal strip and used for screw connections only.

At the lower edge of the photo is a terminal strip designed especially for soldered connections. To the left are two KRE-8 strips, viewed from different angles. These are combinations of solder (pins at bottom) and screw terminal strips. KRE strips are available in different lengths (with 2 to 12 positions), can be soldered onto circuit boards and may be combined with switches, for example, as shown in the photo.

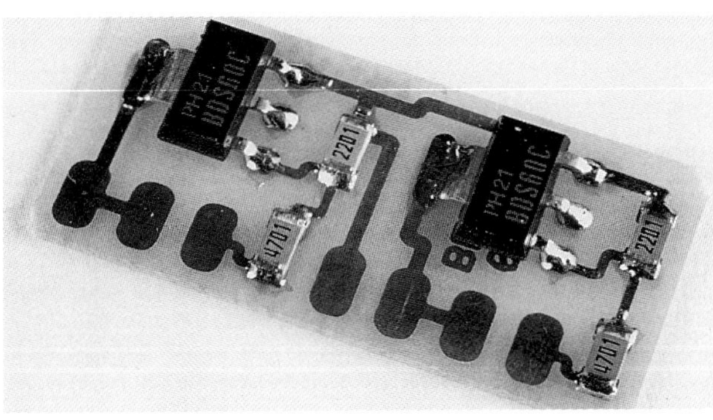

This is an enlargement showing surface components mounted on a circuit board. Don't expect results this good the first time out. Proper tools, patience and some experience are required before results are perfect.

The soldering pencil (as opposed to a large soldering iron) must have a fine and clean tip. Solder should be 0.7 mm or less in diameter and have built-in flux channels.

One is often a little too cautious in applying heat. That's why cold soldered joints are so common on premier projects. On this board, you can see that the solder has spread evenly, creating good contact at all joints.

A third hand for SMD-soldering

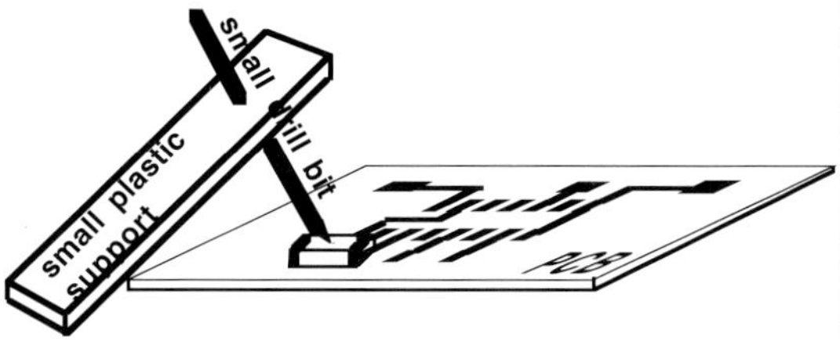

tool called "Eezi-Tweezi" is outstanding for handling the smallest components, such as resistors and capacitors. Regular tweezers or needle-nose pliers can also be used.

To keep components in place during soldering, there are two primary options for the hobbyist. A holding tool (or third hand) can be arranged as shown in the sketch. Secondly, soldering glue in miniature syringes is available. A dab of the soldering glue on the circuit board holds the component in place. Insulating SMD glue is placed midway between the soldering points. Contact soldering glue (that melts down to solder) on the other hand, is placed on the respective pierced solder tags. Heat is then applied with a special heating gun.

An emergency solution: Soldering SMD components

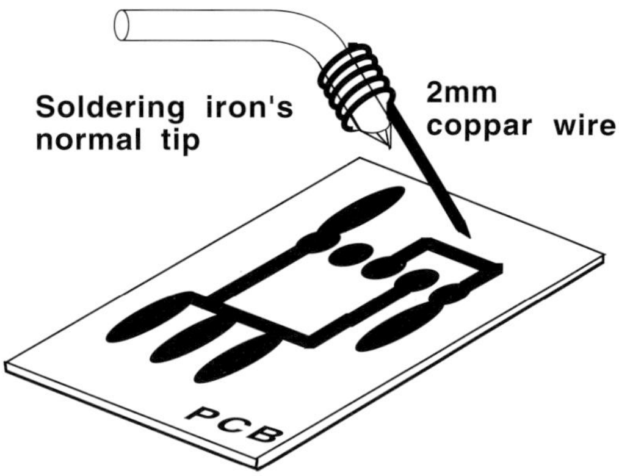

If you plan to solder in a more conventional manner, your choice of solder is very important. Forget all, old heavy solder with separate soldering flux. The solder should consist of 60% tin and 40% lead with channels containing a special, separate soldering flux. The diameter of the solder should be no larger than 0.7 mm.

For successful soldering, you'll need a soldering station,

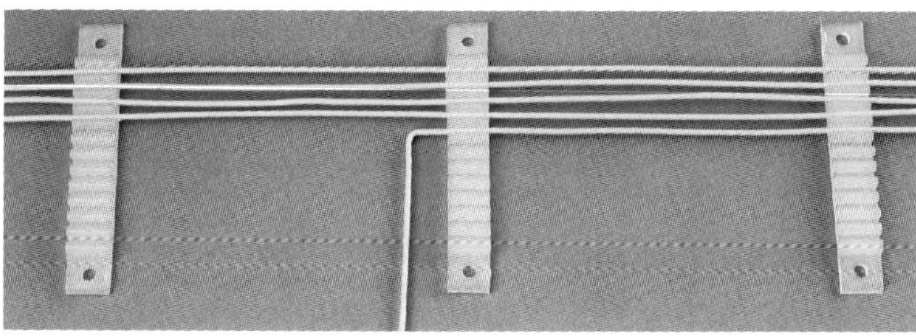

Much of layout construction is electrical. And that is why so many electrical leads are needed for distributing power and signals out to the line. Visitors are seldom urged to look under the benchwork and if you peek under some layouts, you'll see why. Layout wiring often resembles a snake's nest more than anything else. While it is possible to maintain such a layout, it is difficult.

As a workplace, the area under they benchwork is far from ideal. Right from the start of layout construction, you should set things up for tidy and easily maintained wiring while you still have space to work. These plastic holders from Brawa are helpful. Holders are also available in several other configurations from a variety of manufacturers.

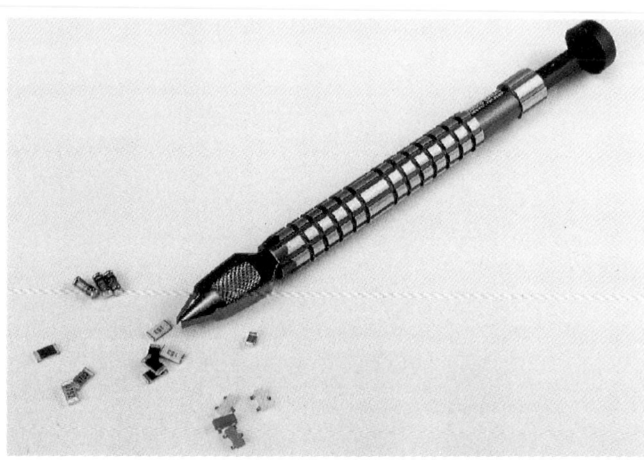

Surface mounted components are extremely small. The hobbyist's entire supply can usually be stored in a matchbox. Instead of needle-nose pliers, you'll need tweezers to handle them.
The photo shows a special plucking tool that elegantly locks around the components. They can then be moved and mounted on the circuit board. A light press of the black plunger releases the grip. Three sizes are available for small, smaller and even smaller surface components.

good lighting and a steady hand. The last item will come with experience. As far as the soldering station goes, you'll need low heat (20-30 Watts) and a fine tip. Soldering stations usually feature interchangeable tips.

When soldering a component in place, the first soldering point is not as important as the second (or with ICs, the following points). As long as only one component pin is attached, the component can be removed by applying heat to just one point. You can check the positioning before going on to the next pin. As soon as a second pin is soldered, adjustment becomes much more troublesome. In view of this, it's always better to begin assembly with the most difficult pin.

A common problem for the beginner is unintentional bridging between two soldering tabs. The gaps are small and it is easy to apply too much solder. The thinner the solder, the easier it is to use the right amount. To correct the problem, you can use a solder removal tool—it sucks up excess solder when heat is applied—or a small solder-braid of thin copper wire.

Another more risk-filled method involves drawing a "line" through the hardened bridge with the iron tip. The solder draws back to the respective soldering points and the short circuit is eliminated. Inspect the area in good lighting with a magnifying glass.

The final phase is troubleshooting and function testing. I usually begin work with SMD modules by soldering the ICs with several pins in place. When this is done, I make measurement checks of the soldered pins. Here it's good to be able to see the copper leads under the IC if several pins are connected together. For this reason, have an extra, unused circuit board or paper copy of the circuit board layout on hand. Measure each pin with a high impedance ohm meter. It is just as important to find short circuits as unconnected contact points. And its just as well to uncover these problems immediately—they won't go away by themselves. You may sometimes need to add a little extra wiring to the module.

If a component must be removed, you can do this by applying heat alternately to the connection points. This will work with 2-pole components such as resistors, diodes and capacitors but not with 3-pole and upward components.

An IC can be removed with a special unsoldering tool. This alternative is both too expensive and complex for the hobbyist. In the rare occasions when removal is needed, it is more economical to discard the components.

ICs can be removed using small clippers or by heating one pin at a time and carefully lifting upwards. The critical point here is that the pin is not lifted before the solder has melted—the circuit board lead can separate from the board surface.

We'll close with this final bit of advice. Although components are heat-sensitive and can be destroyed by too much heat, a more common problem for beginners is the faulty soldered joint caused by under-heating the joint. With time, you'll learn how the solder's structure and surface changes as it melts and spreads. Practice makes perfect.

MINITIP

Substituting parallel with serial resistance

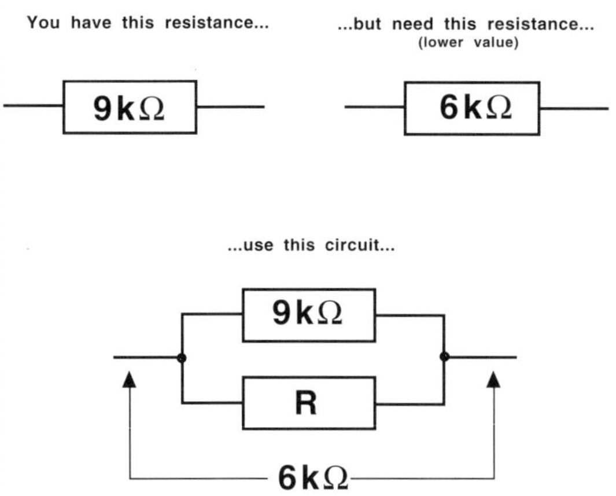

...and calculate parallel resistance as follows...

$$R = \frac{9k\Omega \times 6k\Omega}{9k\Omega - 6k\Omega} = \frac{54}{3} = 18k\Omega$$

Voltage divider

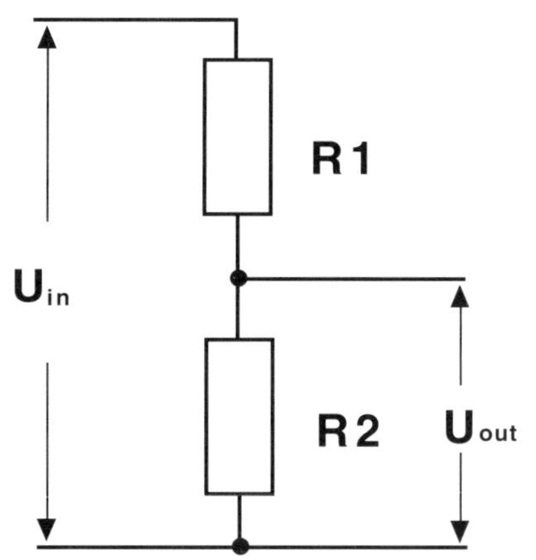

U is voltage in volts
R is resistance in ohms

$$U_{out} = \frac{R2}{R1 + R2} \times U_{in}$$

MINITIP

Full- and half-wave rectifying

Variation of rectification between full- and half-wave is usable for several reasons. When rectifying feed voltages, a higher effect value is attained—and consequently, higher voltage—with the resulting half-wave voltage from a diode bridge.

Another point concerns output frequency. If we assume a 110V, 60 Hz line voltage, half-wave rectification provides 60 Hz at the output. Using full-wave rectification, a doubled frequency of 120 Hz is attained, as shown in the example.

This phenomenon can be used in a controller circuit to run a little slower for switching operations (with half-wave rectification) and a little faster for mainline operations (with full-wave rectification). This is employed in the better controllers featuring pulsed voltage feed to trains.

The illustration shows how to wire a 2-pole switch in the bridge rectifier circuit, made up of four diodes.

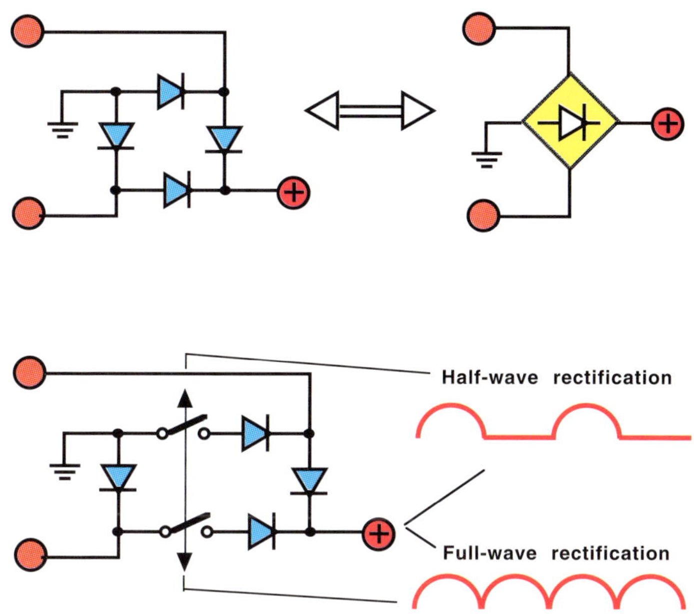

MINITIP

Calculating capacitance

serial connected capacitances reduce total effect

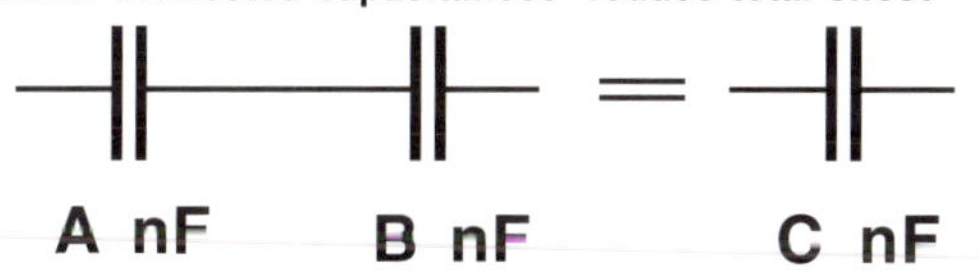

$$\frac{1}{C} = \frac{1}{A} + \frac{1}{B} = \frac{A + B}{A \times B}$$

parallel connected capacitances increase total effect

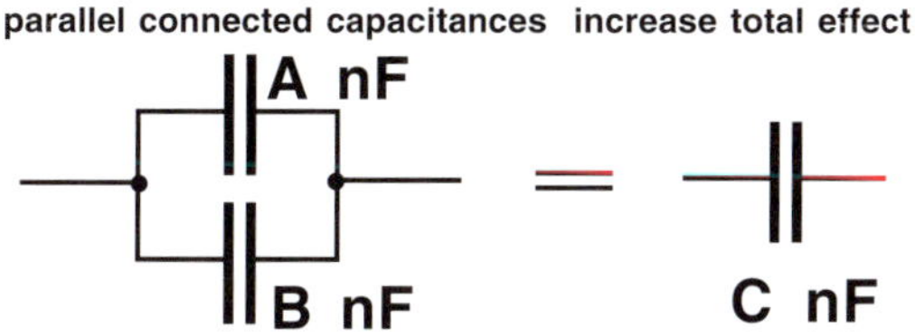

$$C = A + B$$

Making AC calculations

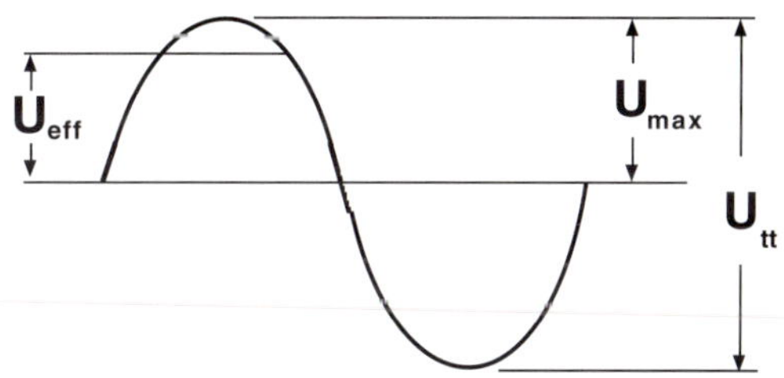

U_{max} maximum voltage in relation to signal ground

U_{eff} effective voltage in relation to signal ground

U_{tt} maximum AC voltage, top to top

$$U_{max} = U_{eff} \times \sqrt{2} = U_{eff} \times 1.414$$

$$U_{eff} = \frac{U_{max}}{\sqrt{2}} = \frac{U_{max}}{1.414}$$

$$U_{tt} = 2 \times U_{max}$$

MINITIP

Warning—High Voltage!

This tip shows how to avoid a serious accident risk. It's a risk that arises from absent mindedness and "clumsy" line voltage connections.

The figure shows two common, approved controllers with built-in transformers for 110V-120V. One has been connected to the wall outlet. The other is unconnected.

In an attempt to increase capacity to my lamps and other accessories on the layout, I have connected the accessory outputs of both controllers in parallel. It sounds logical but introduces unanticipated hazards.

Via the 14V- 16V secondary windings, the lower transformer is *reverse-fed*—110V-120V is present at the plug of the second transformer even though it is not plugged in!

Take this warning seriously and never wire as shown under the "X". What's the solution to the capacity problem? Divide power feed so that one transformer takes care of one part of the layout and the second, the rest.

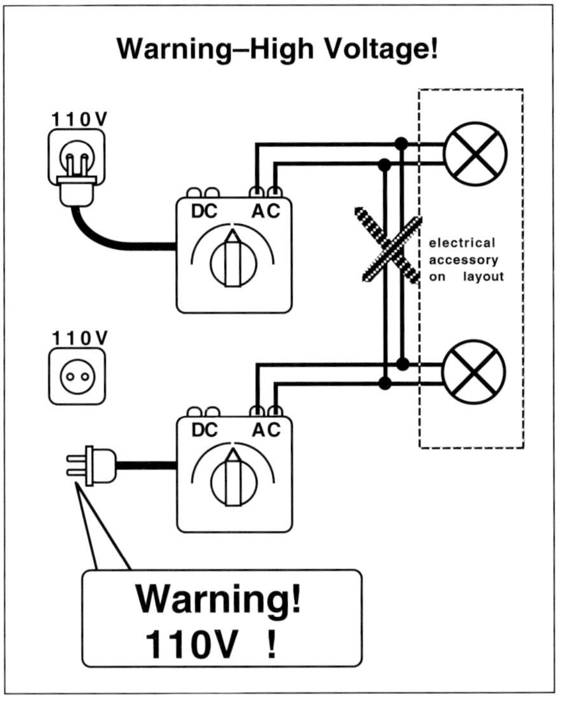

MINITIP

Voltage regulation for AC motors

Some locomotives on the layout are real speed devils—they start with a jump and their low speed registers are nonexistent. This circuit is designed for AC motors. M symbolizes the rotor, field windings and the reversing relay—the components present in a standard, analogue Märklin HO locomotive.

With this circuit, you can reduce the drive voltage drastically for the problem engine and attain entirely different, low speed characteristics. LM 317 is an adjustable voltage regulator. IC1 and IC2 each control their respective half of the sine-formed AC voltage. This means that the respective trimpots must be adjusted so that the secondary voltage to the motor is symmetrical.

It is preferable to make the adjustments with the aid of an oscilloscope.

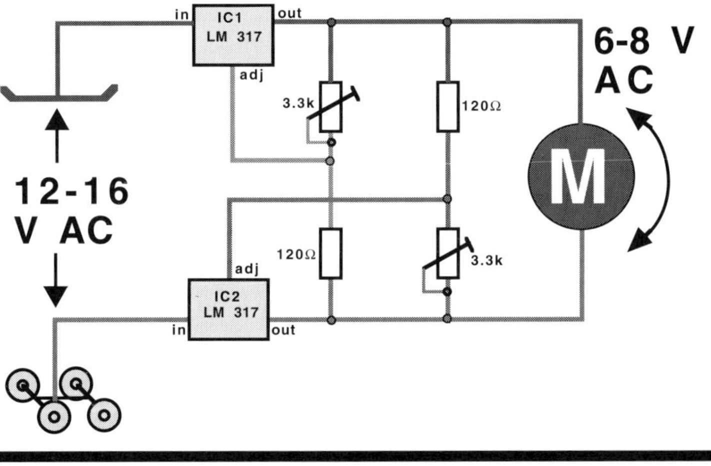

MINITIP

Reducing motor power voltage

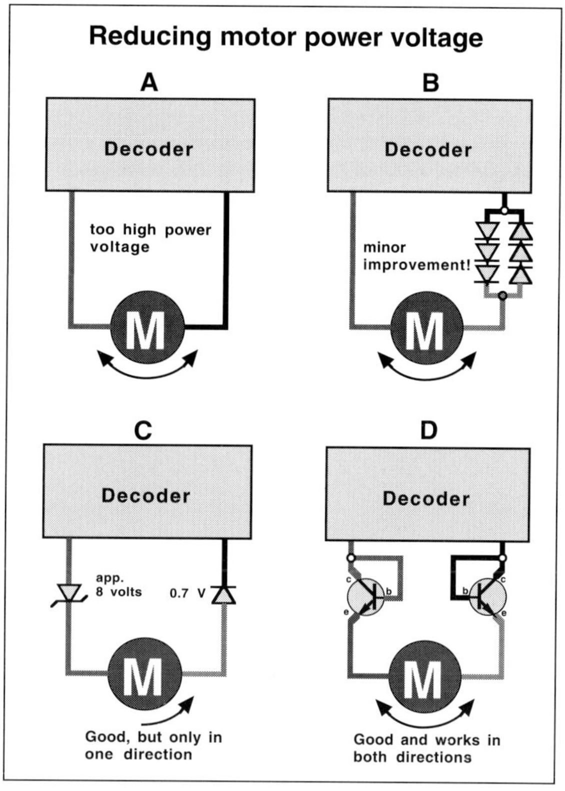

This tip concerns smaller DC motors and how they can be adapted to different voltages. Let's say that you have a voltage source for a motor of approximately 16V but that the motor should be powered with 8-10 volts. The voltage source can be pulse breadth modulated.

With the high voltage, the motor starts with a jerk and runs hard. Most likely, the motor with eventually be damaged by this heavy service. The usual trick is to connect a double row of effect diodes (for example, 1N4001 for currents up to 1A) as shown in Alternative B. This gives us a reduction of just over 2V, or rather, 0.7V per diode. Diodes are needed in both directions so that the motor can run both forward and backward.

Alternative C shows a circuit permitting a single rotation direction only. The combined "effect" of the Zener-diode and effect diode is 8-9 volts and the motor now runs at a more agreeable 8 volts.

From an electrical standpoint, Alternative D is the equal of Alternative C, except that it works for both rotation directions. Transistors connected in this manner function as diodes with 0.7V voltage drops. In the blocking direction, the drop is approximately 8 volts. Naturally, one must be sure to dimension the transistors so that they handle the heat loss (effect in Watts) of 8V times the current needed by the motor.

As an example, I used a 6V motor (270mA) with two BC337 transistors and a total running voltage of 15 volts. It worked fine.

MINITIP

Common electrical formulas

Ohm's and Watt's laws are the most common formulas encountered by the electronic hobbyist. The two formulas express the relationships between voltage (U) in volts, current (I) in amperes, resistance (R) in ohms and effect (P) in watts.

With the accompanying wheel chart, you can find the desired quantity when two of the others are known. In some formulas, quantities are squared and in others, the square root must be calculated. For example, the square root of 16 is 4. Conversely, 4 squared is equal to 16. I squared (I^2) is the same as I x I, or current times current.

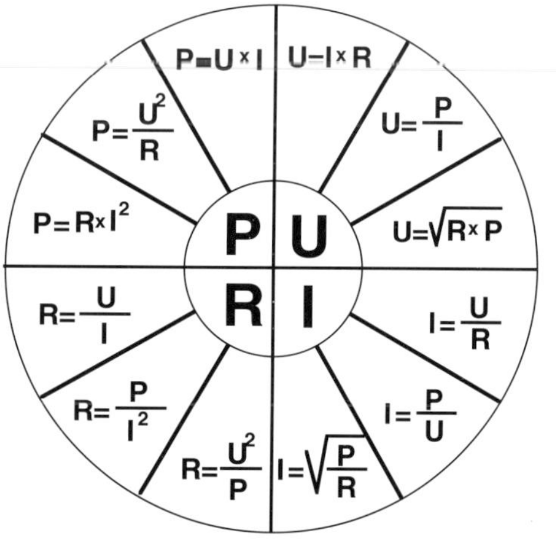

MINITIP

Voltage converter for low-voltage motors

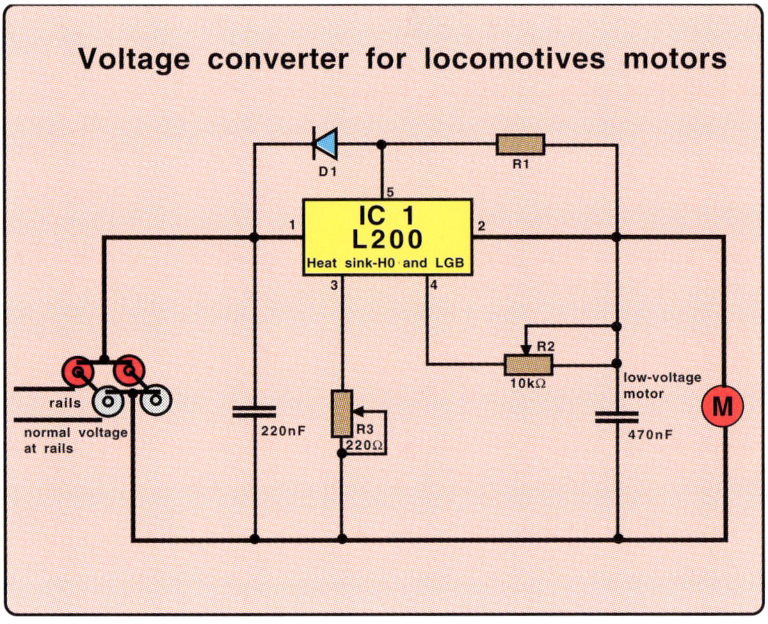

You may have a locomotive that is just too quick. Or maybe you would like to add a motor to a locomotive or car but the voltage at the rails is too high. If so, this module can help.

When an engine is running in its primary direction of travel, current enters IC1 at pin 1, exits at pin 2, passes through the motor and then returns to the tracks. Maximum speed can be adjusted with R2 and current limitation with R1. R3 sets the working voltage.

For N scale, D1 is a 1N4001 diode and R1, a 0.47Ω/1.5W resister. This limits the current to a maximum of 950 mA. For HO, D1 consists of two parallel wired 1N4001 diodes and R1 is at 0.33Ω/2W for a maximum current of 1.4A. Finally, for LGB, D1 is a 1N5401 and R1, 0.22Ω/4W for an output current of 2 ampere. Current limitation begins when a voltage difference of 450 millivolts is reached between IC1's pins 2 and 5.

The L200 circuit is thermally fused against overloads and triggers at approximately 100°C (212°F).

MINITIP

Reduce voltage with diode bridge rectifiers

The bridge rectifier is usually associated with half-wave AC rectification. This is its prime duty. In this circuit, the rectifier is used to drop voltages and save space.

When one needs to reduce the operating voltage for a motor, for example, it is often suggested that a chain of parallel, counter-wired diodes be used. You'll see this technique often in this book. There is a 0.7 drop in voltage over each diode. Note that counter-wiring is a must for bi-directional current flow.

To reduce an engine's voltage by 1.4 volts, you would need four diodes, connected as shown. If you use a "short circuited" bridge rectifier instead, your component total drops from four to one.

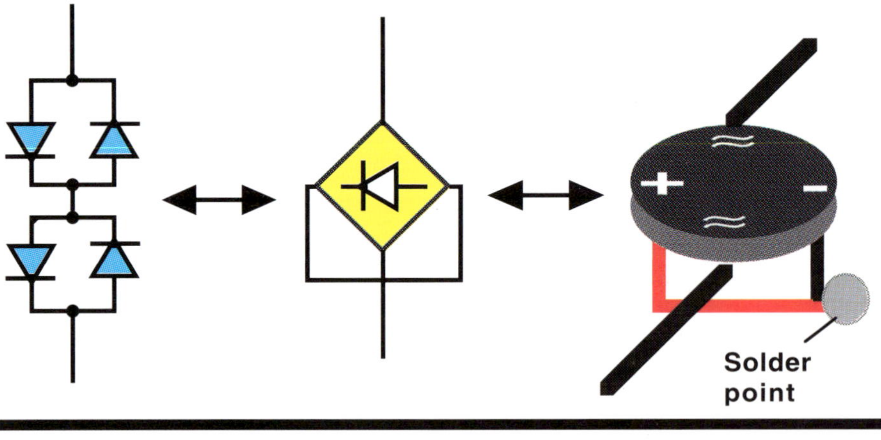

MINITIP

Logic probe for TTL and CMOS

The multipurpose instrument and logic probe are the instruments that I use most. With a logic probe, one can see logic levels and pulses of various types.

The logic probe presented here is my absolute favorite and I have probably built five or six of them—sometimes it's good to be able to connect several to a circuit. A close friend is the brain behind this module and I shall be perpetually grateful to him for coming up with such a useful probe.

All six inverters are contained within a single IC and the entire circuit will fit in a larger, ball-point pen shell. The test tip is made from heavy, solderable copper wire. The two probe cables, attached to plus and minus, are made from the most pliable copper wire obtainable and equipped with alligator clips. To use the probe, just find the appropriate test points and connect the respective clips. The probe can be powered with 5 volts for TTL or 3-15 volts for CMOS.

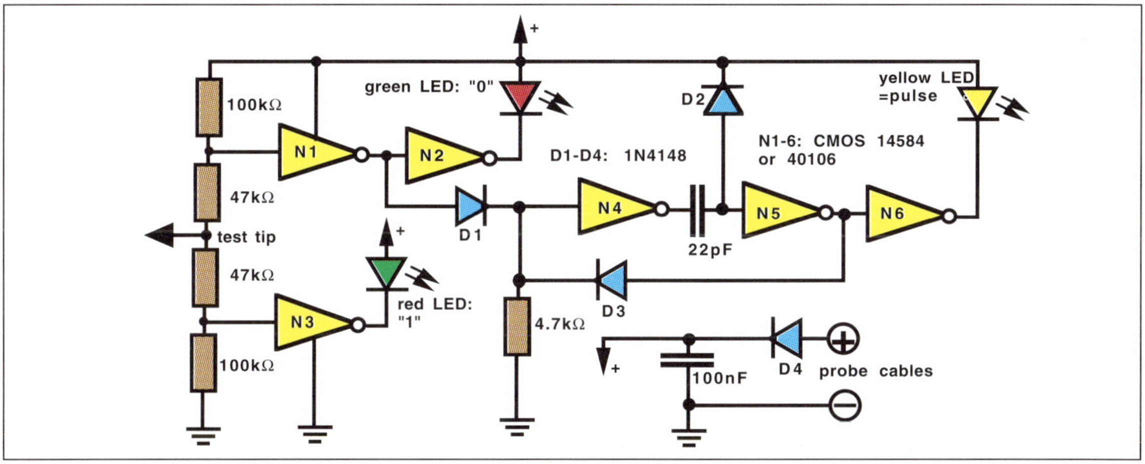

MINITIP

Simple switching controller

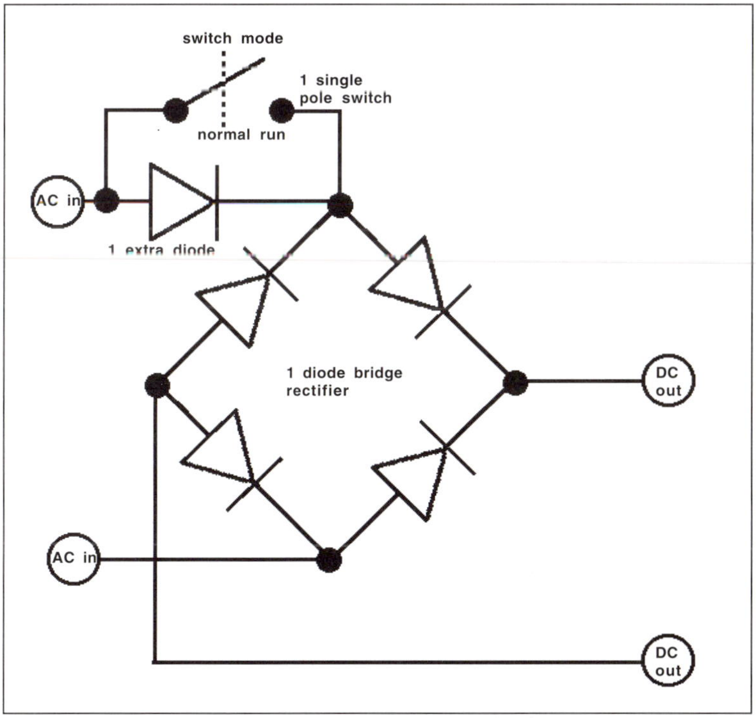

There is a separate minitip that explains the principles behind full- and half-wave rectification. There, four diodes and a 2-pole switch were used.

This circuit is a variation of the same theme, but with cheaper components. We utilize a complete diode bridge rectifier and by switching a separate diode (see the figure), we attain the same functionality as before. This time, a single pole switch—they're less expensive than their 2-pole brethren—is used.

It should be noted that current in this version passes through three diodes instead of two, as with half-wave rectification.

Digital Construction Projects for 2-Rail Systems (DCC)

There are several different digital control systems for model railroads and it isn't all that easy to know which one is being referred to. But there are two systems that are in common use. These are the two that I will cover in this book.

One of the systems is from Märklin and is normally used for AC Märklin HO rolling stock running on special track featuring a third rail. Since the '80s, many of us refer to this system as, "3-Rail digital", or just plain, "3R digital". You can read more on the system in another section of this book. In some documentation you will also find references to "AC-digital" and the "Motorola format" which also focuses the Märklin Digital system.

For the moment, we'll take a closer look at the 2R digital system—the control system that is compatible with nearly all 2-rail, DC products. All scales are represented in this group—from Z scale (1:220, Märklin Miniclub) to scale 1 and LGB (1:22.5), often called G-scale.

The 2 rail digital system is today commonly known as NMRA-DCC, and the projects in this section work with the DCC protocol.

RUN ENGINES DIGITALLY WITHOUT DECODERS

The originator of DCC is a German, Bernd Lenz from Giessen. Lenz came up with the ingenious idea of transmitting digital pulses with power via the rails to the locomotive, regardless of which direction the loco was placed on the track. He has a patent on the invention. He also found a way to *run a locomotive without a decoder* using the same digital system. He has a patent on this too!

Lenz' consulting firm was so successful with development of this system that both Arnold and Märklin bought in to the system. Since the late '80s, they have periodically marketed it with their other model railroad products—Arnold for N scale; and until 1992 Märklin used it for their own system and scales, as well as independent systems for DC operations.

2-RAIL DE FACTO WORLD STANDARD?

Until the Fall of 1991, interest in digital technology for model railroads was rather tepid in the US. At that time, certain forces within the world's largest model railroad organization began to move in the direction of digital multiple control. The NMRA (National Model Railroad Association) and its technical section established a committee under the lead of Stanley R. Ames with the objective of creating a standard for digital control of model railroads. For some unfathomable reason, I was invited to join this committee. Imagine; I had never even been a member of the NMRA, before this event.

We worked intensively for two and a half years to create the basis for this standard. In August 1993, NMRA's board of directors made the first approving decision (two consecutive approvals are required for ratification of a new standard) on the new standard which for the most part was compatible with what Lenz had developed. The final approval was made in April 1994 and the new standard presented at the NMRA convention in August of the same year.

This is a first prototype of a 2R decoder. It can be assembled in a few hours by someone with a little experience. Prior to the prototype phase, a number of experimental circuits were made to verify that the circuit, and its sub-circuits, worked as intended. During the brainstorming phase, most work was on paper, in the form of sketches and calculations.

A few of the projects built for the 2R series. Shown are NMRAF 4, 6 and 7. After the prototype phase, as seen on the previous page, comes a second drawing phase for printed circuit board design. Early design work is done on paper while the final work is done on film using special rub-on profiles. A scale of 2:1 is used for hole components and 4:1 for surface mounted components. In the future, some circuit layouts will be prepared on the computer.

MANY HAVE ALREADY ADOPTED DCC

The European model railroad manufacturers have shown much interest in adopting the American standard as an intentional standard of sorts. On the European market, systems from Arnold and Märklin are already available. In addition, Lenz Elektronik GmbH has its own DCC system, marketed under the name of Digital-plus. Märklin has later decided to stick to one system (Motorola) and left DCC in mid 90's. However through the purchase of Trix (January 1997) they have channels both to the Selectrix and DCC systems.

An American manufacturer, Digitrax, offers complete DCC systems in accordance with the NMRA standard using entirely US-made components. Other US electronics companies have also announced production of DCC products.

This is a very interesting development. It means a great deal for model railroaders. More manufacturers on the scene means more media coverage, as well as more interest from other companies that don't want to "miss the boat". This creates competition and, hopefully, lower prices. And thanks to the standard, one can continue with products from another supplier if the first should go out of business or otherwise terminate manufacture of DCC products.

BACKGROUND TO THE NMRAF SERIES

During my time spent on the NMRA committee, I have had reason to make a number of experiments to verify different transfer protocols, extra functions, etc. Some of these were so successful that I have gathered them in the following section under the collective name, NMRAF Series; that's to say, NMRA Function modules.

In several cases, circuit board layouts are provided as a guide if you would like to make your own. Experience has shown that it is often difficult to obtain good film reproductions when copying from paper prints. For this reason, work is underway in collaboration with an independent company to try to help in distributing film reproductions (and possibly, even circuit boards). If this effort bears fruit, you'll find information in the book's introduction. The same is true for software, that may possibly be distributed in a similar manner. Please also look for updated info at "http://www.hobby.se/Rutger/Rutger.html".

READ UP ON SURFACE MOUNTING

Finally, I would like to point out that some of these modules are built with SMD components (Surface Mount Device). This is absolutely necessary for digital control of Z, N and HO layouts. Other projects can be built with the more conventional components, where the most common ICs are in DIL (Dual In Line) casings.

If you have not worked with surface components previously, I would recommend a glance at the section on tools and soldering. In the section, I have described the techniques common in the world of miniature components.

NMRAF 0 – universal control module

Decoders in N and HO scales have often but one function output—that for headlights. If control of additional functions such as extra lamps, sound or simpler animation is desired, NMRAF 0 is one alternative.

The European decoders in the above mentioned scales are built around a microchip from Phillips programmed by the German electronics firm Lenz. Check the IC labeling on your decoder. To be able to use the decoder for these projects, it should be labeled, "LE 10" (the first version) or "LDE" followed by two digits (LDE 20 or higher). As of this writing, the latest LDE version was 33. There is a version 4x from Lenz but in a casing with too tight pinning for soldering.

On all of these decoders, function signals F1-F4 are available but not "wired" for the user. This control module will make them operational.

This isn't much of a locomotive but it runs and can be remote controlled. A little extra ballast was needed to improve operational reliability. This is a miniature, rolling lab where the 0 module was installed and function-tested.

SELECT YOUR COMPONENTS

The basic variant—NMRAF 0—is built with surface components (SMD). If you would prefer working with hole-mounted components, there is a slightly larger module, NMRAF 0.1 that uses these somewhat larger components. The illustrations and component lists are provided for both variants.

A little advice on soldering SMDs can be in order. The smallness of these components constitutes a threshold for many in beginning with SMD technique. I take it for granted that you have read up on SMD mounting in the book's introduction, so I'll put basic skills aside for now.

Before starting a project such as this, I first make a dry run and do a bit of planning. How will I solder the various components? How much working space do I need around an IC? Two general pieces of advice are to solder components with several connections early on, as long as this doesn't hinder mounting of other components; and to solder low-profile components in place before those with higher profiles.

SOLDERING WORK

In this case, it is natural to begin with the IC and to then select an "adjustment" pin. As long as I have just one soldering point affixed to the IC, I can easily adjust its position. For this reason, I begin with a *difficult soldering point*—one that will be difficult or impossible to get at later. In general, the pins out at the corners are easier to solder than those in the middle. And double pins—two pins to the same soldering tab—are easier than singles. Consequently, I select a single pin in the middle, pin 4 for example, and apply a *little* solder on the soldering tab.

I then position the IC and melt pin 4 into the solder. Take a look at the other pins and their soldering tabs. Does it look right or should the IC be moved a hair? If it needs to be moved, heat pin 4 and move and/or twist the IC slightly. Repeat the operation until you are satisfied. As soon as a second pin is soldered it will be difficult, if not impossible, to move the IC.

Continue with the resistors and transistors. These do not interfere with each other because of their similar component profile heights. Use the same soldering method as before with solder applied to a single soldering tab. Heat the component into position and make adjustments as needed by heating this single point.

"PIGGYBACK" MOUNTS

There was a thought behind my layout design for the NMRAF 0. The module fits back-to-back with the previously mentioned decoders. Attach them to one another as shown in the illustration with double-sided tape to reduce

How to wire NMRAF 0

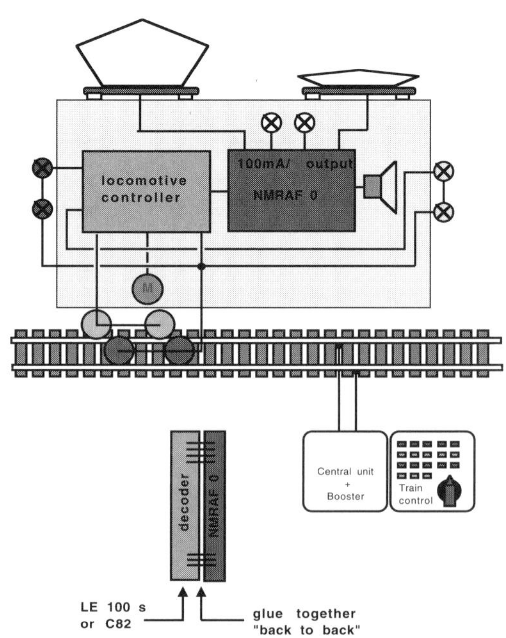

How to wire the decoder

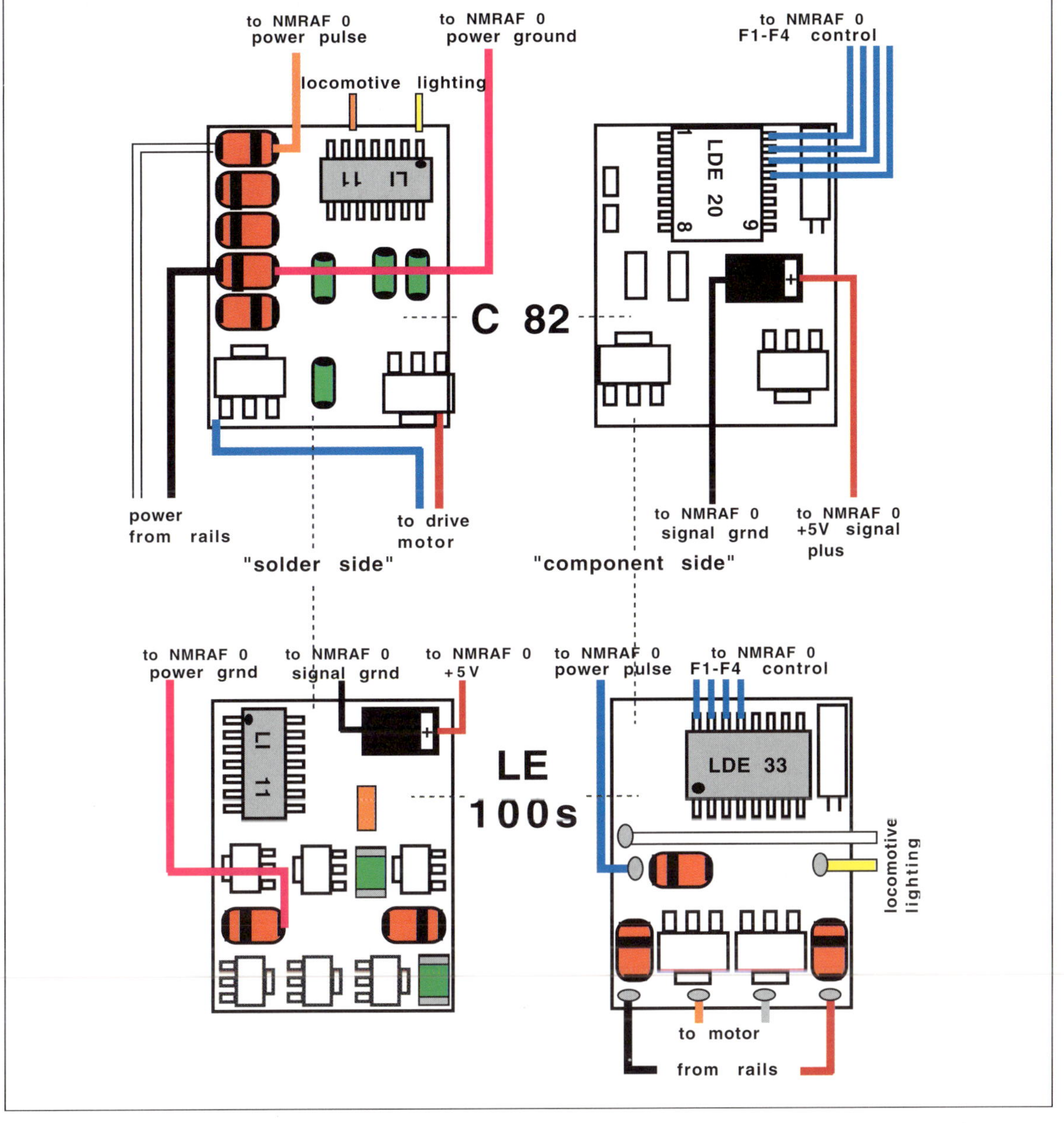

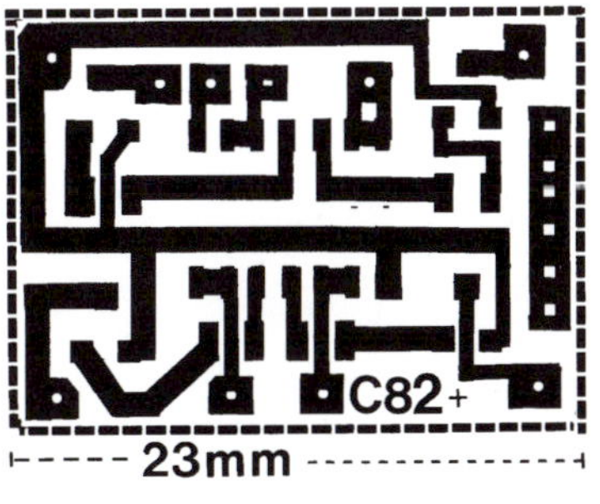

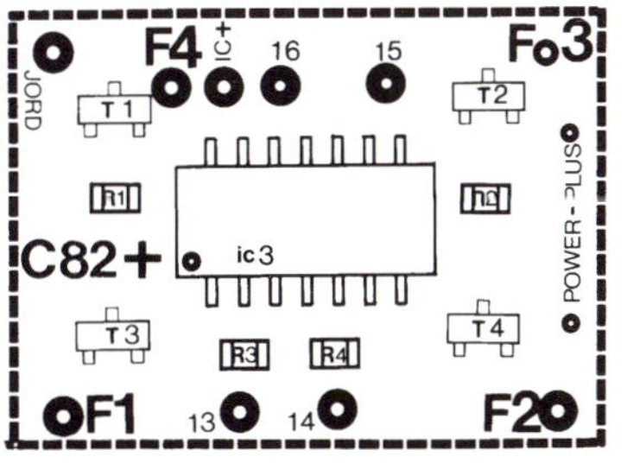

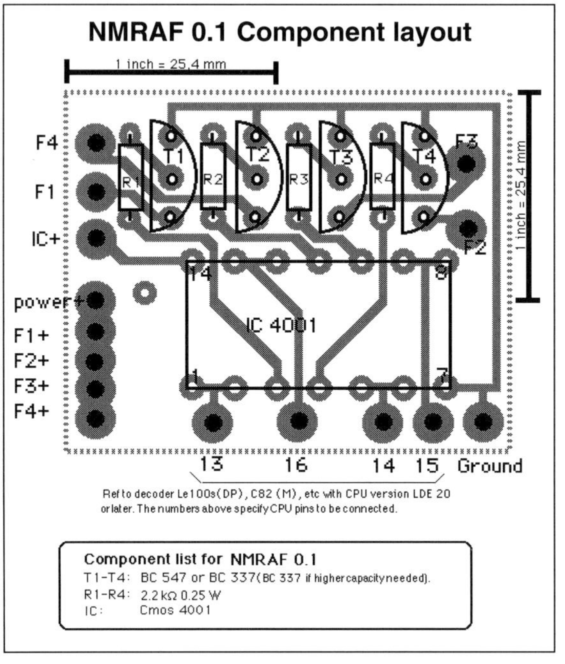

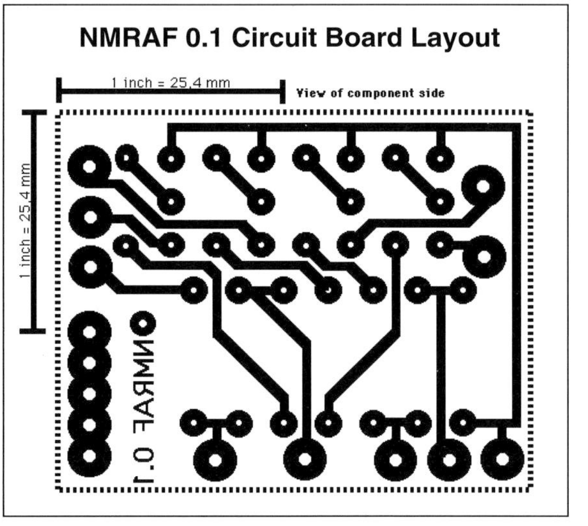

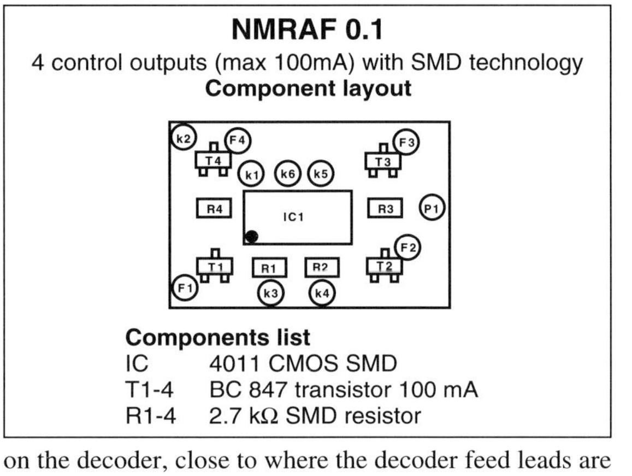

lead lengths.

On the decoder's LDE IC, we will connect pins 13-16 as control inputs to our module (k3-k6). A five volt signal plus for operation of IC 1 (k1 connection) can be found on the decoder adjacent to the LDE IC, at the boxy, black electrolytic capacitor. One of the capacitor's leads goes to +5V and the other, to ground. Make a check metering if you have the possibility to connect the connecting leads to a power voltage. No other wires need be connected for this test.

Note that labeling on small surface mounted electrolytics can be deceptive. Several manufacturers use a dash to label the anode (+). On common electrolytics, we are accustomed to a dash (or minus sign) being used to designate the cathode (-).

Ground and power plus—the feed used to power the proposed accessory—is accessed at the rectifying bridge on the decoder, close to where the decoder feed leads are connected. The rectifier is often discretely constructed (with four diodes in a row). See the accompanying figure. Finally, terminals F1-F4 on the module (see the component layout) are the connections for lamps and other accessories that you would like to control.

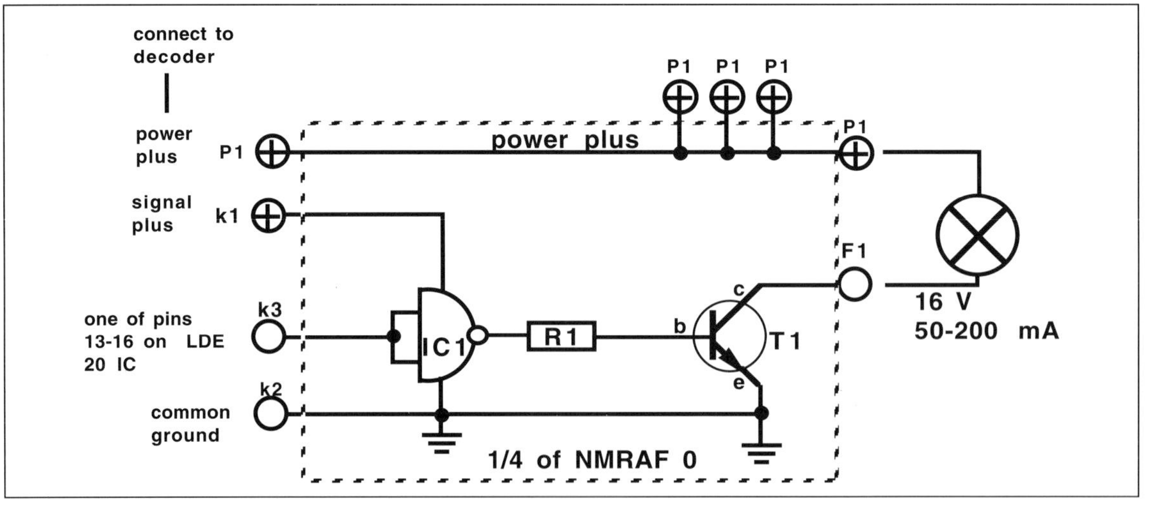

NMRAF 1
Locomotive decoder for all scales

This module has just three components. Naturally, it is not really a complete decoder but can rather be seen as the brain of a decoder. The onboard components are the three most important and they are required for several of the locomotive decoders. One can connect components to the module that vary depending upon drive capacity, scale, function requirements, etc.

The thought behind NMRAF 1 is that one can experiment with 2R digital technique at a low cost and develop a locomotive decoder that corresponds to your own needs.

Block diagram for 2R digital engine decoder

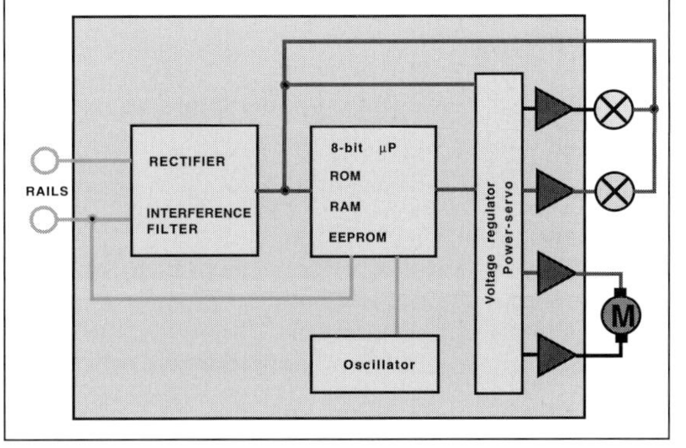

THREE COMPONENTS AND ONE CIRCUIT BOARD

The small microprocessor, IC 1, is the central component for several of these projects. Its frequency is controlled by a small 4.19 MHz crystal located on the circuit board.

IC 2 is a small monolithic special circuit developed by Lenz expressly for use as a locomotive decoder. It contains, among other things, a 5 V stabilizing circuit for the CPU (Central Processing Unit, or microprocessor) power feed and motor pulse control. Actually, IC 1 has a built-in power transistor that can control a small N scale motor directly without an external H-bridge. Normally though, we use the circuit shown here with four transistors and four diodes connected around the motor in an "H".

The device provides PWM pulses (Pulse Width Modulation). This means that the motor is powered with short voltage pulses for slow motor-spin and longer pulses for

MINITIP

H-bridge for scale 1 and LGB

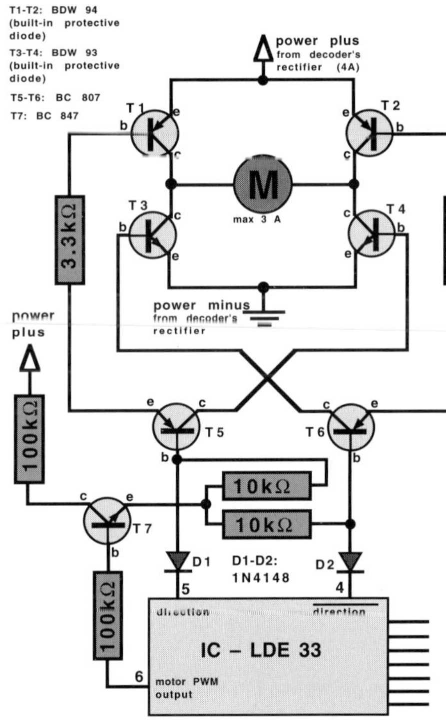

H-bridges are described at several other places in this book. This variant is intended for larger locomotives with their more powerful motors, Consequently, there are differences from the other H-bridge circuits.

The four hefty power transistors are of the Darlington type with built-in protective diodes. This means that complementary , transient diodes are not required between, ground, the motor and plus.

Furthermore, the H-bridge is powered via two transistors because base currents to the power transistors are greater. T5 and T6 are alternately driven (depending upon motor direction) by output 4 or 5, directly at the decoder's main IC (LDE 33). Via D1 or D2, the base is lowered to zero for one of the two transistors, regardless whether PWM pulses are present at the motor output (pin 6) transistor T7. Note that pins 4 and 5 are never low (equal to zero volts) simultaneously.

By power pulse, we mean the feed from the rails via an appropriately dimensioned diode bridge. Preferably, the diode bridge should be common for the entire decoder. Alternatively, use a ground for the diode bridge that is the same as for the decoder's diode bridge.

quicker spin. It should be mentioned that motors with ironless rotors, such as Faulhabers, should not be run with PWM for extended periods. Use an analog, filtered (from pulses) voltage instead.

Our device does not differ from other decoders and problems should not be exaggerated.

DIRECT CONNECTION OF HEADLIGHTS

Component capacities are dependent upon the scale we are building the decoder for—different scales require different component capacities. If we stick to N or HO scales, 50 mA bulbs are usually appropriate for headlights. These can be connected directly to the outputs on IC 2. If larger loads will be powered, use an external power transistor (a BC 337 with a 300 mA capacity, for example) with the transistor base connected to IC 2 via a 2 kΩ resistor.

A FEW PRACTICAL TIPS

It's worthwhile giving some thought to soldering work before starting construction of the NMRAF 1. Considering that the ICs are angled towards each other, you shouldn't begin with the large IC 1, as is customary. Start with the pin on IC 2 that is most difficult to solder. I usually start with pin 6 and center the circuit around pin 6's soldering tab. The order in which you solder the remaining pins is unimportant. Be sure to examine each connection in good light so that any short-circuiting solder "pearls" can be removed.

If you want to be absolutely certain that everything is right, use your multi-purpose instrument to check the resistance between the pins and soldering tabs. I'll give you an example of how to make these checks. Pins 9 and 10 on IC 2 are soldered to a common, double soldering tab and that is connected to pin 7 (also on IC 2). The copper lead continues on to IC 1's pin 8 and the ground connection. First locate the ground connection (connection pin #4 on the NMRAF 1 component board). Connect one instrument test lead to the ground connection and step through the circuit just described.

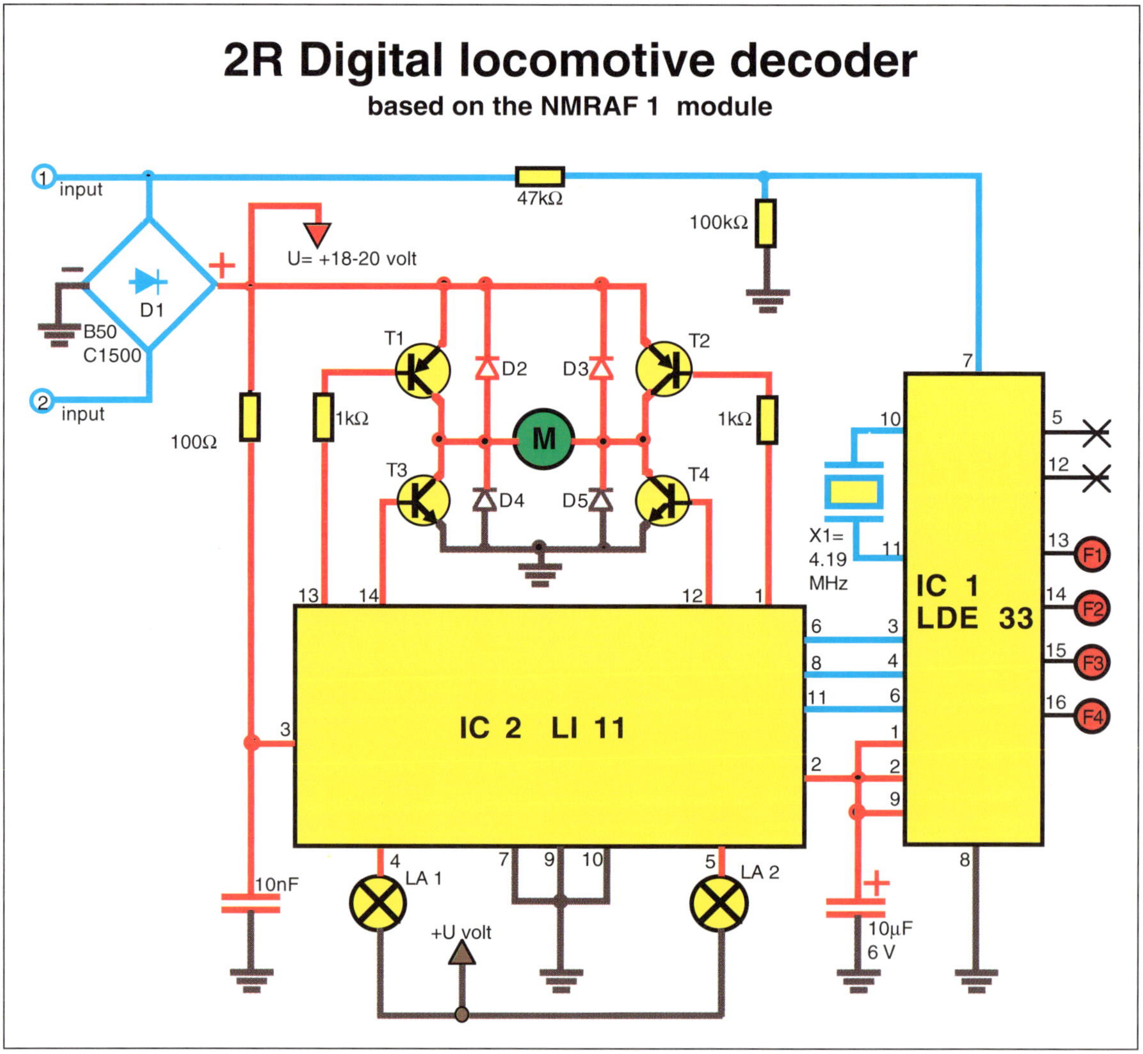

2R Digital locomotive decoder

based on the NMRAF 1 module
(circuit board connections)

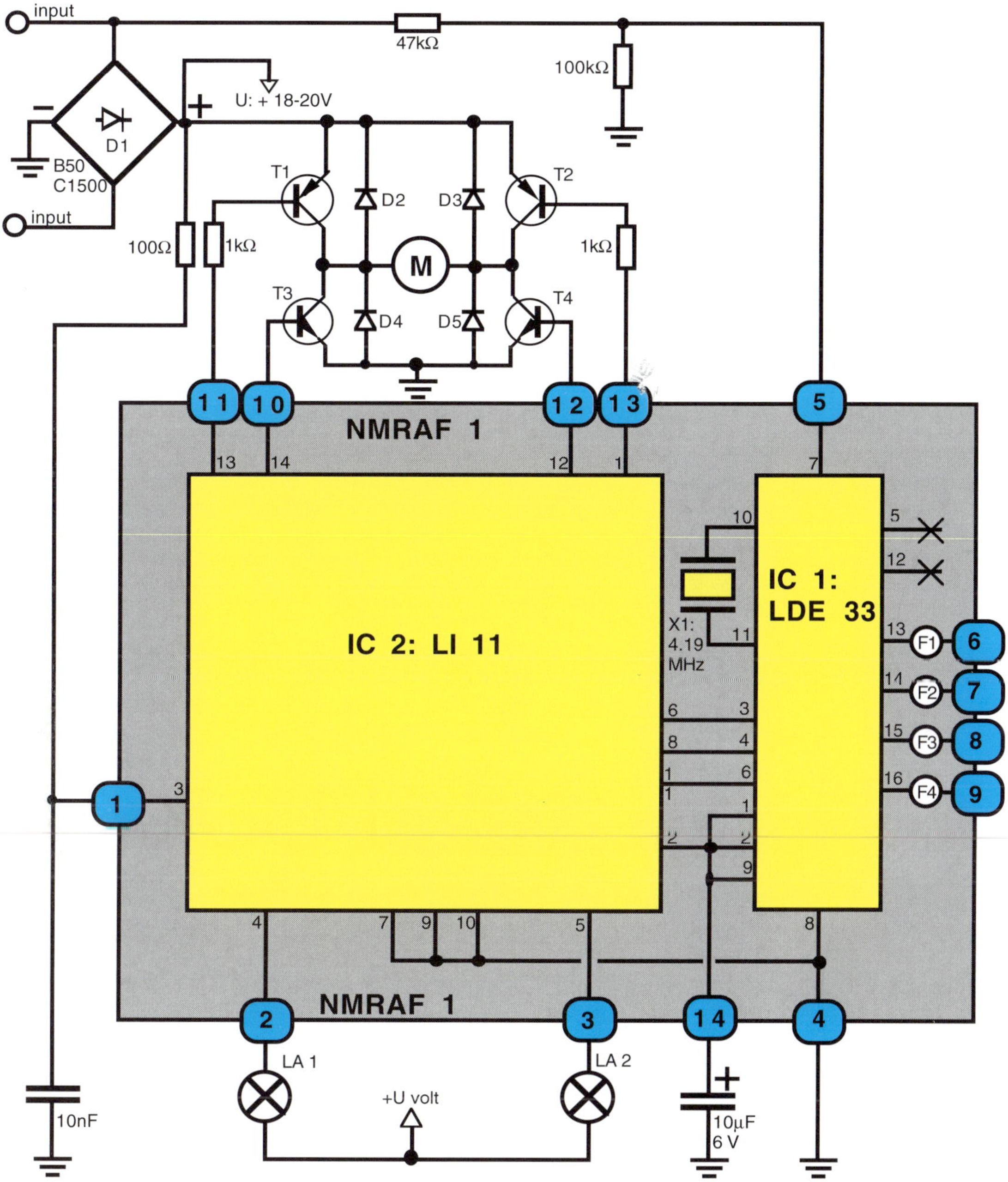

GOOD AND BAD SOLDERING

Begin by testing just *above* IC 2's pin 7. If the soldering is right, the instrument should indicate zero ohms (0Ω). A corresponding measurement at pins 5 and 6 should indicate no connection (insulated). If this measurement (or the measurement for pins 8 and 11) indicates no resistance, an inadvertent bridging between connection points has occurred. So pull out your magnifying glass and a knife with a thin, sharp blade. This problem must be straightened out if the module is to function properly.

Another item that deserves a close check is the lead running from IC 2's pin 2 to IC 1's pins 1 and 2. It snakes between pins 10 and 11 on IC 2 and must be checked if you want to be sure that everything is OK. The check involves seeing that the lead is intact (check the soldered connection for pin 2 on IC 2) and that there is no excess, short-circuiting solder at the double soldering tab for pins 9 and 10 on IC 2.

In practice, inspection takes much less time than it takes to read through these instructions but I hope you now understand what I mean with after-soldering checks. It will save time in troubleshooting if you know that the basic blocks have already been double checked.

GO EASY ON THE SOLDER

It is not for reasons of economy that I urge you to spare the solder. It is just because it is easy to use too much solder and difficult to remove it later. You should use fine gauge solder with several flux channels. I always use Ersin Multicore Solder. It is 0.7 mm in diameter with a mixture of 60% tin and 40% lead and has numerous flux channels. Frugality with solder begins when we attach IC 1 to the first

MINITIP

Remote controlled coach lighting

In conjunction with the construction of NMRAF 1, inclusion of a brief tip could be in order. The locomotive decoder can naturally be used for much more than control of locomotives and other motorized objects.

This minitip will cover construction of an inexpensive remote control module that can be used to control interior lighting for passenger cars. One could give each coach its own unique address to allow simulation of a conductor making his way through a night train, switching on coach lighting one car at a time. Or assign all coaches the same address as "their" locomotive —the choice is yours.

Programming of the decoder is the same as for normal locomotive decoders with one exception. Despite correct in-programming of the new address (or addresses), the new address is not displayed on the programming unit. This is because the coaches lack motors. Instead, try giving the command for on and off of the new address as a function control.

Note that total lighting current may not exceed 100 mA on the unmodified circuit. This means that you can use a maximum of three 30 mA lamps or two 50 mA lamps. Use an external power transistor for larger loads.

Remote controlled coach interior lightning with NMRAF 1

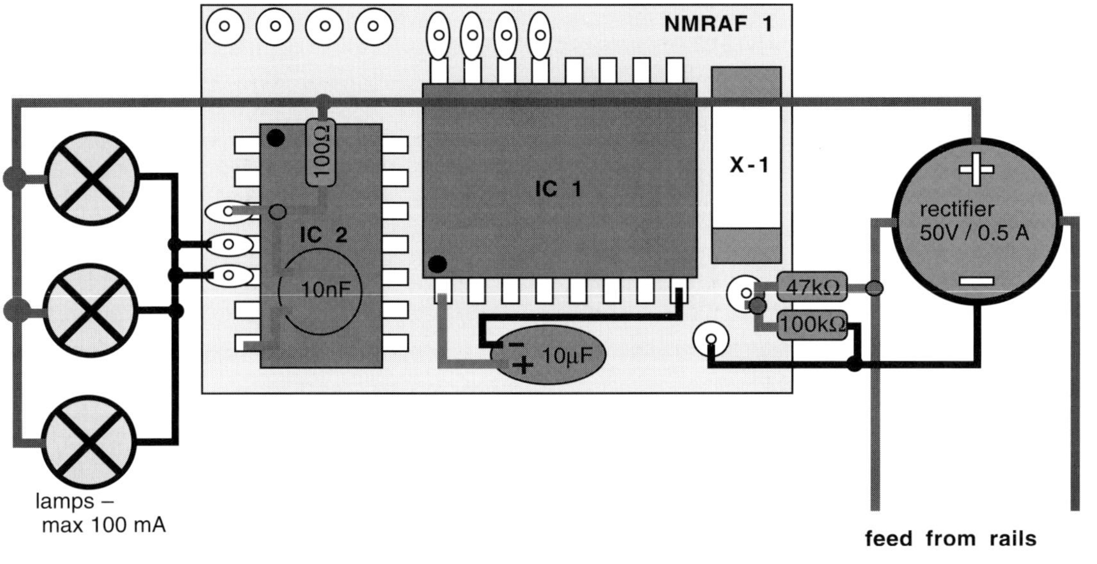

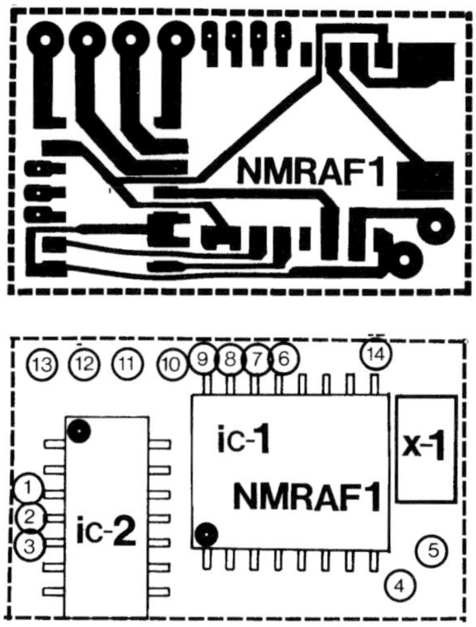

NMRAF 1: Components list

IC 1	LDE 33 (or earlier version) CPU
IC 2	LI 11 servo-IC
X1	X-tal 4.19 MHz

Locomotive decoder: Components list

1	NMRAF 1
C1	cap 10 nF
C2	cap 10 µF / 6V
R1	resistor 47kΩ
R2	resistor 100Ω
R3	resistor 1kΩ
R4	resistor 100kΩ
LA1-2	lamp 16V / 50 mA
D1	diode bridge B50C1500
D2-5	diode 1N4148
T1-2	transistor BCP 68
T3-4	transistor BCP 69

soldering tab. I begin with pin 3 as a centering point and make sure that pins 3, 10 and 11 are soldered at an early stage. These are a little more difficult to solder than the others.

Lastly, the crystal is soldered into place to complete the NMRAF module. For the sake of clarity, I'll go into detail on soldering connection point 14 (see the component layout) in that it does not have a soldering tab with a center hole as have the other connection points. The intention is that a 10 µF electrolyte capacitor should be used here. A tantalum electrolyte would be appropriate in that these capacitors take so little space. One can then use connection point 14 as a lead from pin 9 on IC 1. Placement-wise, it can be better to solder the capacitor's anode to pins 1 and 2 on IC 1. Electrically, this common connection point has the same potential. Consequently, the capacitor's other lead (the cathode) connection point is close by in that it is connected to ground.

MINITIP

H-bridge for HO scale

H-bridges are circuits used for motor control. The components are placed around the motor in a configuration that resembles the letter H.

In the minitip for the corresponding construction for N scale, you will find a description of the circuit's function. Let me add here that the circuit makes use of pulse width modulation (PWM) to change motor speed.

All out-pulses to the motor are at full drive voltage but at varying lengths of time. Short pulses at long intervals provides slow speed and long pulses at short intervals, high speed. In the interval between pulses, drive voltage is zero volts.

PWM frequency is typically 100-200 Hz (polarity swings per second).

If you would like to construct a compact locomotive detector, the bridge can be combined with NMRAF 1, as is described on these pages. Use SMD transistors and double-diode components (see L 21 in the component descriptions). Note that the IC LI 11 is included in the NMRAF 1.

H-bridge for HO-engines

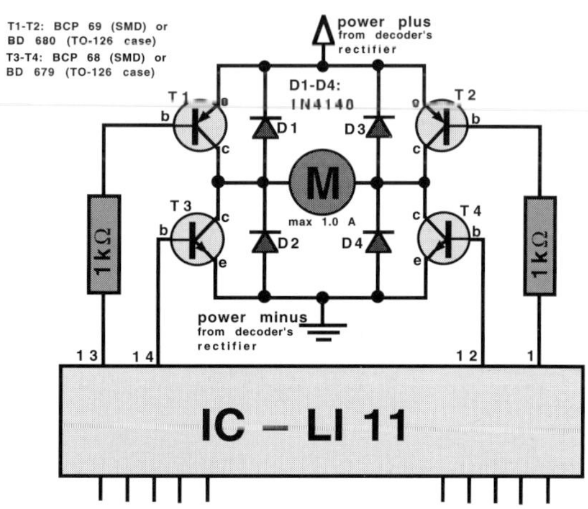

NMRAF 2 – Switch Decoder

This decoder is small enough to be mounted directly on switches. It can be glued to the switch to form a single unit. The circuit board measures just (34 mm x 23 mm). Surface mounted components have been used to make it as compact as possible,.

A distinguishing characteristic of the NMRAF 2 is that it is controlled as a locomotive decoder and not as an accessory detector. The reason is as follows. If you have a smaller layout, with just a few locomotives and switches, obtaining a separate keyboard for switches can be an unjustified expense. It is advantageous to use the train controller for the switches too. After all, there are at least 99 addresses available. This should be sufficient for both the layout's locomotives and switches.

Another case where this type of decoder is invaluable is when operating with Märklin's infra red system. The IR controllers can only (in their original configuration) call locomotive addresses. If you want to throw a switch, you have to dash back to the control panel. This defeats the purpose of wireless control. Using the NMRAF 2, even switches can be controlled from these handheld units (they resemble television remote controls).

CONTROL WITH FUNCTION KEYS

Switches are thrown with keys F1 and F2 on the controller. There is one important thing to remember—you must double-click the keys! The reason for this is that we need momentary activation only. The first click activates the extra function and the second, deactivates it. This differs from other keyboard functions and there is nothing we can do about it.

Double-clicking is vital because current flows through one of the switch's coils until the function is disabled. Modern switches have an end-position breaker and current is interrupted automatically, should you forget the second click. But if the switch is later thrown in the other direction, an "electrical conflict" arises in that both coils are activated. You'll notice the conflict by the buzzing from the switch. If you should forget to double-click an older switch, it will quickly overheat.

If you are using an Arnold or Märklin Control 80 control unit, a red lamp clearly indicates when an extra function is active. On the Digital Plus Cab LH100 active functions are indicated by marked digits on the LCD-display.

ALTERNATIVE SUB-ADDRESS

The switch is normally controlled via F1 and F2 but you may also use keys F3 and F4. Take a look at the circuit board layout and especially, pins 13-16 on IC 1. As standard, pin 13 is connected to resistor R2 (this is F1) and pin 14 to R1 (this is F2). Under the IC, the alternative is provided to select either pin 15 or 16 for F2. If this is what you want, clip the copper lead at pin 14 and make a soldering bridge to the desired pin. I would suggest pin 16

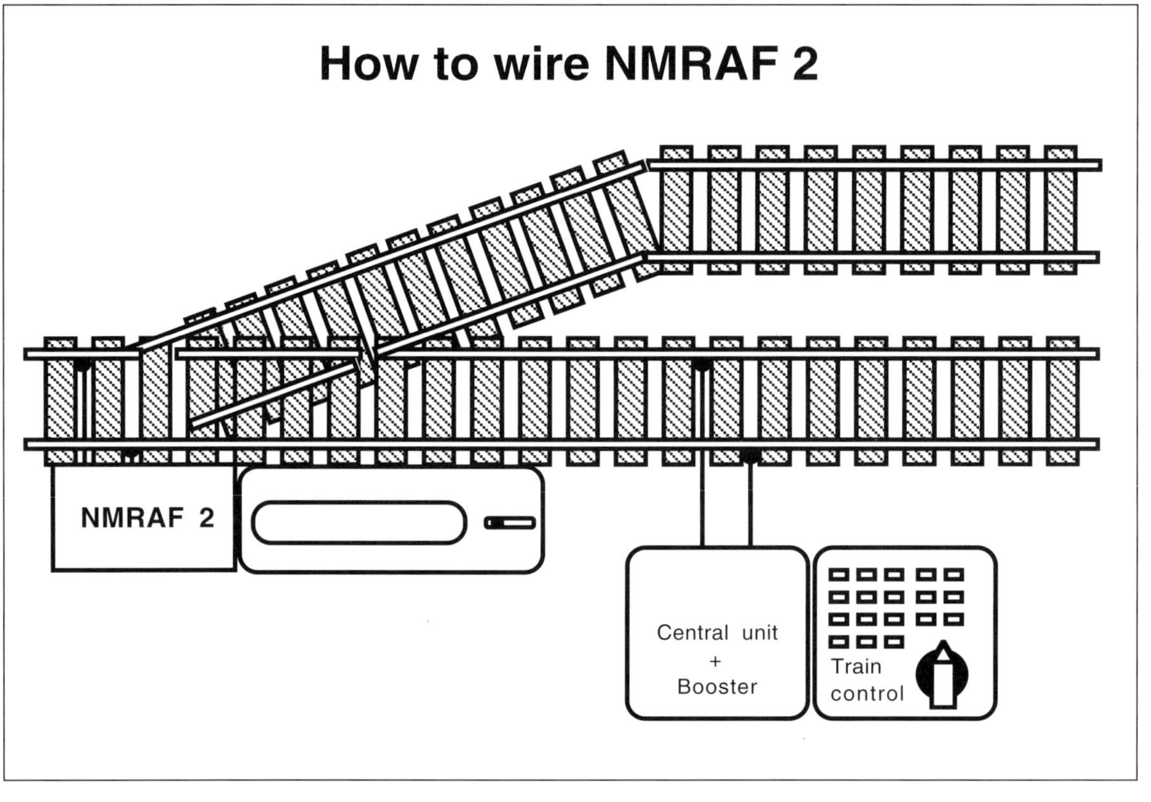

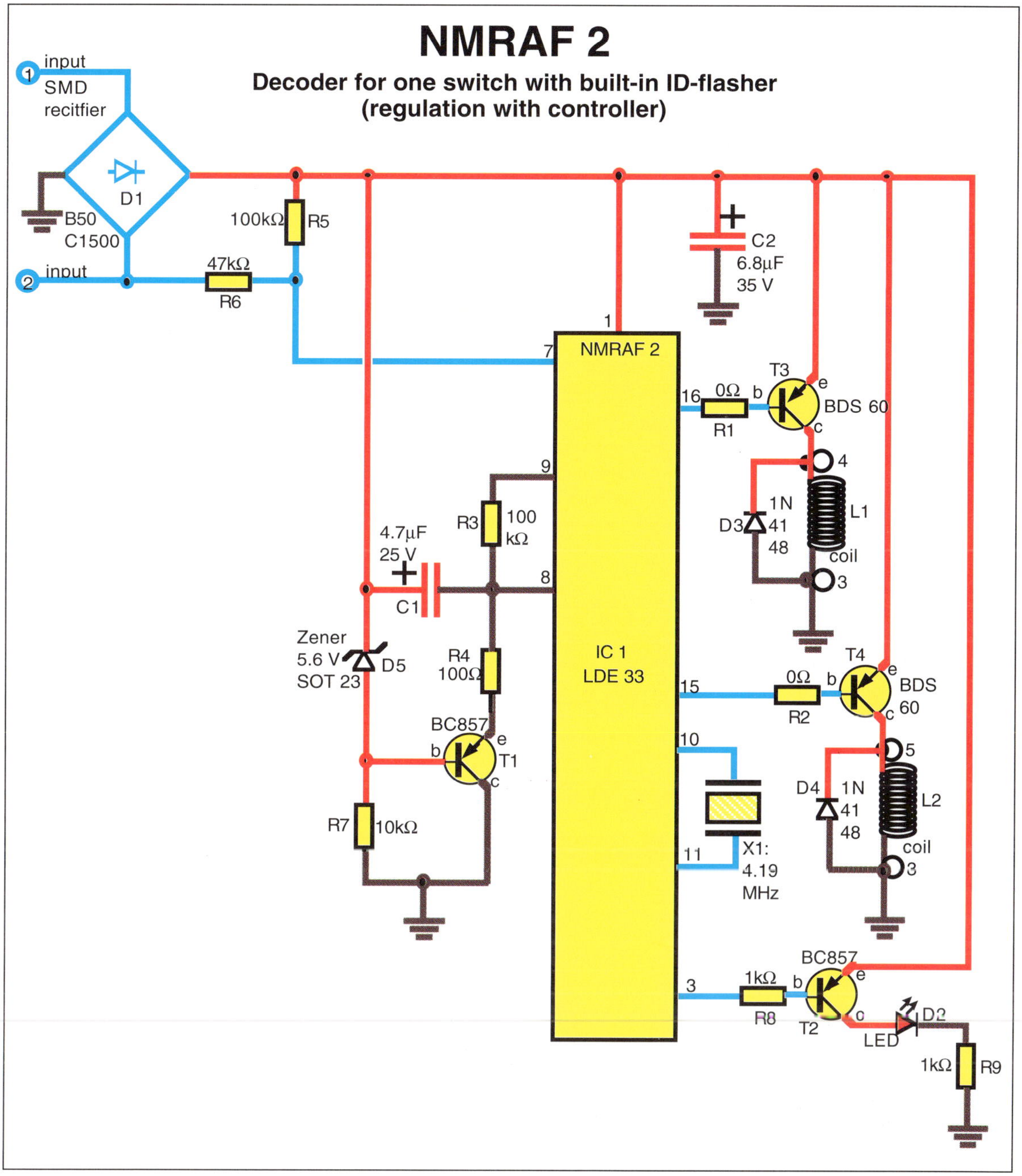

(this is F4). In a corresponding manner, the copper lead for pin 13 can be clipped. Bridge between R2 and pin 15. Because this shall be done before soldering IC 1 in place, pin 15 can be straightened and soldered directly to R2.

SMART ID CONTROL

The problem: You have several adjoining switches in a yard and have trouble keeping track of their identifying numbers. Throwing the wrong switch can be disastrous if it happens to be under a speeding train.

The solution: Each NMRAF 2 decoder features a small LED that can be activated *before* throwing the switch. F0 (normally used for locomotive lighting) is used to turn on the LED. This provides immediate switch identification, giving you the opportunity to avoid serious problems.

The LED is visible even when covered with ballast. If you prefer installing the decoders under the layout, the surface mounted LED can be replaced with a common LED, lead up to the layout surface through a drilled hole. If you would rather do without this indication feature

This turnout is entirely self-sufficient. The only connections are to the rails for electrical power and signal data. NMRAF 2 can be mounted to the side of the turnout as shown, or beneath. One of the module's finesses is its built-in LED ID control. This allows us to check that the right unit was selected BEFORE throwing the switch. See the main text for more information.

entirely, omit all components from pin 3 on IC 1.

CHANGE OF MAIN ADDRESS

Even though the decoder is programmed in the same way as a common locomotive decoder, no indication is provided on the programming unit because no motor is attached. Instead, use the F0 on/off command at the decoder address as a function indicator.

A FEW ASSEMBLY TIPS

It is best to begin with IC 1. I start with pin 3 for centering reasons. After this, I install the following components, in the order listed: R2, T4, R1, T3, R5 and R6. This takes care of "ground zero" and I can now work around the flanks. Save the larger components, D1, C1 and C2 for last.

While many components can be installed in any direction, this is not the case for D3 and D4 or C1 and C2. I have encountered bridge rectifiers with plus and minus positioned diagonally to one another, rather than the usual adjoining-positioning. Positioning is marked on the casing. If you have a rectifier that does not match the setup shown in the circuit board figure, lift the plus pin a tad and run a short lead to the proper connection point.

TESTING AND TROUBLESHOOTING

If this is your first SMD module, it can be good to check your work in accordance with the methods described in the section covering NMRAF 1. Firstly, you need to know that current flows from "A to B" for all conductors on the board. Secondly, you must be sure that there are no inadvertent bridges between adjoining conductors. When you are through with this, you can make a test connection.

Normally, IC 1 is programmed with address 01 at delivery. This is not always the case. I have come across new decoders with addresses 03, 25, 38, etc. To avoid searching, you can program the decoder directly and select the desired final address.

I recommend connecting the decoder via a few short leads, directly to the central unit's (or booster's) outputs. This eliminates any sources of connection error. Enter the correct address and check that F0 turns on the LED. If it does, connect a switch and test operation.

If the LED does not light, the LED could be defective. This is very rare. Check if you have connected the LED incorrectly. Although the wrong address may have been selected or programmed, there can also be problems with the module itself.

LOOK FOR SIMPLE FAULTS FIRST

If there is a short circuit in the decoder, you'll know soon enough. The central unit will detect the overload and shutdown power. The pilot light will also turn off. You can

NMRAF 2 - components list

D1	diode bridge SMD 50V/ 1.5A
D2	LED red SMD type SOT 23
D3-4	diode 1N4148 SMD
D5	zener diode 5.6 V type SOT 23
T1-2	transistor BC 857 PNP SMD
T3-4	transistor BDS 60 PNP SMD
C1	cap 4.7µF/ 25 V tantal
C2	cap 6.8 µF/35 V tantal
R1-2	resistor SMD 0Ω (noll)
R3	resistor SMD 100kΩ
R4	resistor SMD 100Ω
R5	resistor SMD 100kΩ
R6	resistor SMD 47kΩ
R7	resistor SMD 10 kΩ
R8-9	resistor SMD 1kΩ
IC 1	LDE 33 CPU fr Lenz
X1	X-tal 4.19 MHz
L1-2	Switch motor coils
1 st	pcb NMRAF 2

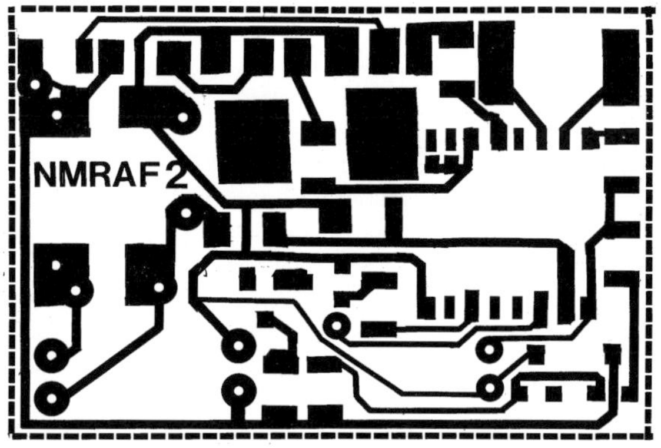

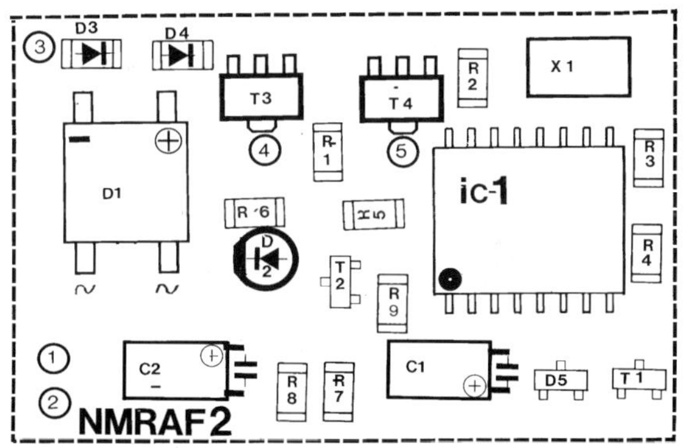

assume that there is an unintended solder bridge on the module.

If the module does respond to a correct command, you can connect two lamps in place of the switch coils. It may be that something in the ID circuit with the red LED is damaged and that the decoder is functional in all other respects. If this is the case, the lamps should light when the appropriate function key is pressed—have you selected F3 and F4? If no lamp lights, something could still be wrong with address selection but there can also be problems with D3 or D4. On occasion, I have found diodes incorrectly marked. Try removing them. As long as the lamps are connected, they are not needed. Their purpose is to shield the respective transistors from transients when the magnet coils are activated.

DON'T FORGET TO CHECK THE FEED

I assume that you have already checked that the voltage feed reaches IC 1. Between pin 1 (plus) and pin 8, you should have approximately 5 volts. If full track power is present, namely 16-20 volts, a serious assembly fault has occurred and it is likely that IC 1 is defective. This means that pin 8 has in some way come in contact with the module's signal ground, which is the same as minus on the bridge rectifier.

On the other hand, if the check shows zero volts, there is no power feed to the IC. If there is full power at the rectifier's plus and minus connections, the fault is with on of the components T1, D5, C1, R4 or R7.

TIME FOR THE OSCILLOSCOPE

If the correct voltage is present at IC 1, use an oscilloscope for advanced troubleshooting , concentrating on the IC 1 pins. Use pin 8 as a ground reference for the oscilloscope. The pulse code comes from the rails at pin 7. Both pins 8 and 9 shall have a constant ground potential. Pin 1 should have a constant feed voltage. All function pins 13-16 shall be high while in the standby state (a logic level of one, corresponding to the voltage at pin 1). This also applies to pin 3. Finally, check the voltage at pins 10 and 11, where the crystal is connected, With a good oscilloscope you should be able to detect a *low* voltage at a high frequency (4.19 MHz).

If all is well up to this point, focus on pin 3. There, with the correct address activated on the central unit (and F0 depressed), you should be able to see the logic level switch from high to low, that's to say, ground potential.

If you have come this far and the module is still not reacting, there are just two main alternatives remaining—either the IC is defective or the intended address are not stored. In the first case, replace the IC. In the second case, perform a reference programming with a locomotive to the desired address. Could it be that something is wrong with the programming unit?

NMRAF 3
Alternative Switch Decoder

Both NMRAF 2 and 3 are designed for connection to, and control of, switches. There are a few important differences between the two. We shall now concentrate on NMRAF 3, which is controlled by a keyboard on the control panel. This is the most common control method for larger layouts and other stationary electrical accessories. Note that NMRAF 2 is maneuvered from the train controller unit and its address is a locomotive address. Here we are working with switch addresses and we can use up to 256 switches in the present version of the 2R system.

As opposed to NMRAF 2, NMRAF 3 has a built-in programming function that can even be controlled remotely from the keyboard or the control panel function.

LARGER COMPONENTS ARE EASIER TO SOLDER

In contrast to NMRAF 2, several thruhole components are used on NMRAF 3. IC 1, for example, is a 20 pin integrated circuit that is easy to work with. The same is true for the bridge rectifier.

There is no onboard ID check as featured on the NMRAF 2. You'll just have to get by without it.

A SOMEWHAT UNUSUAL CONSTRUCTION METHOD

The single sided circuit board has a common soldering and component side. Soldering IC 1 to the soldering side may feel odd. Also unorthodox is the mixture of surface components and hole components. The reason for keeping one side completely free from components and conductors is so that the module will be as thin as possible and easy to glue to a flat surface.

Begin by soldering IC 1 in place, followed by D1 (on the soldering side). Note that in this application, D1 shall not be pushed all the way down but rather positioned so as to permit access with the soldering pencil tip. Continue with the SMD components in whatever order you prefer.

PROGRAMMING VIA THE KEYBOARD

When you need to change addresses, the process is quite different from that used for locomotive decoders such as NMRAF 2. No separate programming equipment is needed. Instead, we use a process involving both the keyboard and

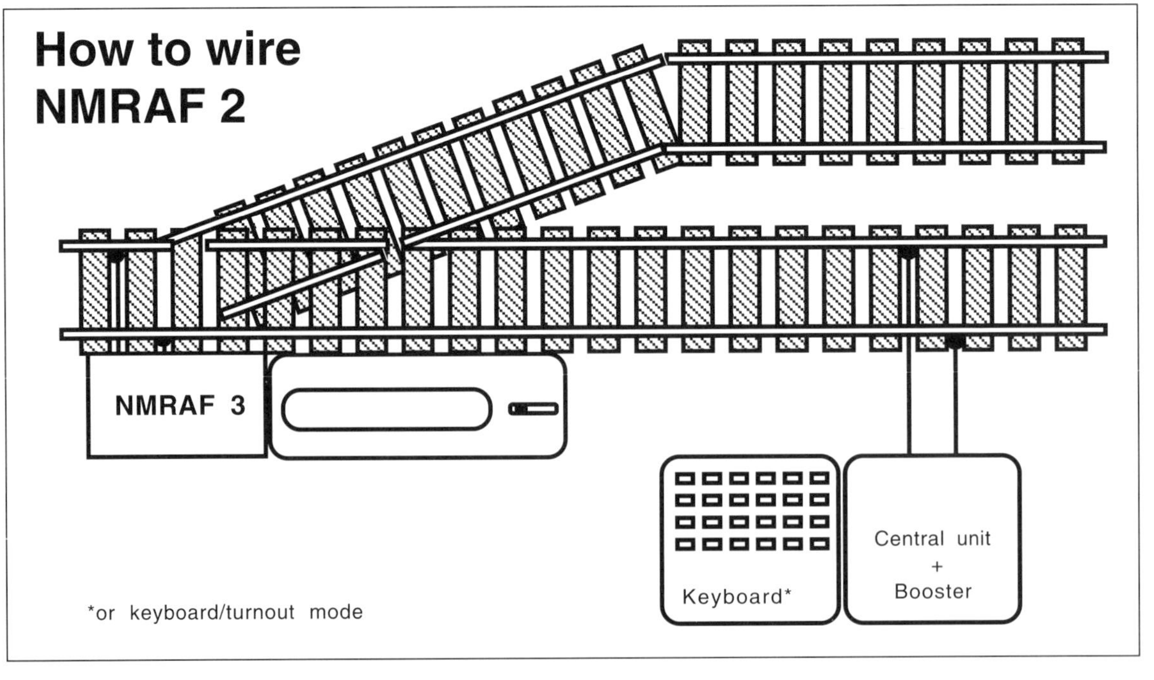

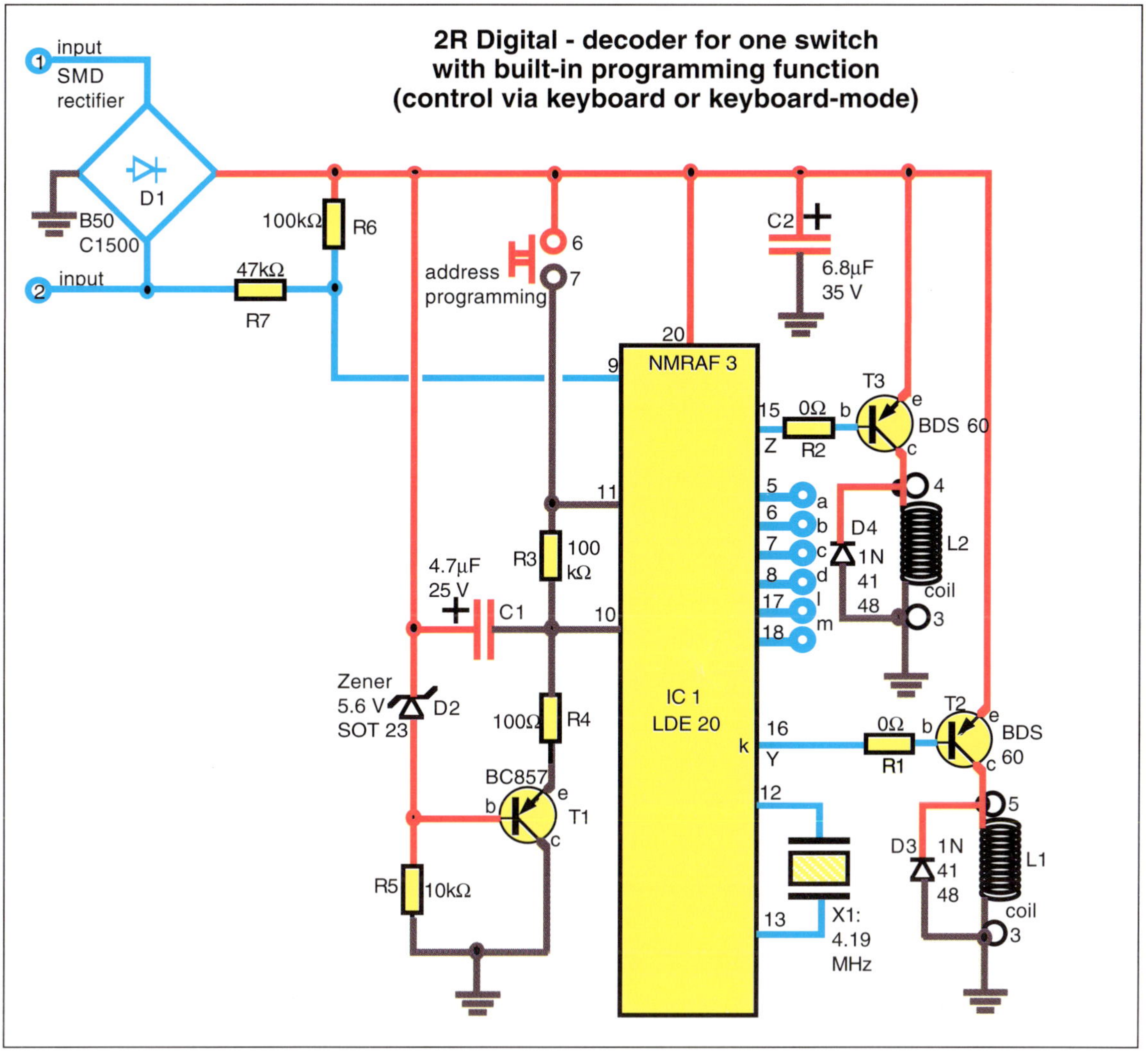

the decoder module itself.

Switch decoder addresses are always in groups of four. This means that the first four turnout addresses are included within the same decoder address. They are identified within this address by sub-addresses. One reason for this is that a decoder chip can control up to four double functions (such as four switches) using just one address code.

When programming a chip, only the group address is coded. The sub-address is determined by the output pin selected (for more information, see the component descriptions at the end of the book). During the programming sequence, a key is to be pressed on the keyboard unit. It doesn't matter which of the keys is pressed within the group address.

...AND A CUSTOM PROGRAMMING BUTTON FOR EACH DECODER

On the component layout figure are two connection points, labeled 6 and 7. These are to be connected to one another while programming. While this connection can be made with a screwdriver, it is both safer and easier to install a push-button.

The process involves pressing the decoder's programming button while powering up the NMRAF 3. The power connection should be the same as for the central unit or booster, as applicable. The first command sent out to the "line" is the address for the switch decoder. This means that one of the eight keys is pushed within one of the groups. This information is received by the decoder and the value stored, making the unit ready for service.

While pressing the key on the keyboard, the decoder's programming button must be depressed. The final command is made with the release of this button. Repeat the process for new addresses. The reason for all this is that the IC contains a program that checks if points 6 and 7 are connected to one another at power-up. If this is the case, the program jumps to the location, where the next incoming code is stored as a group address (there are a total of 64

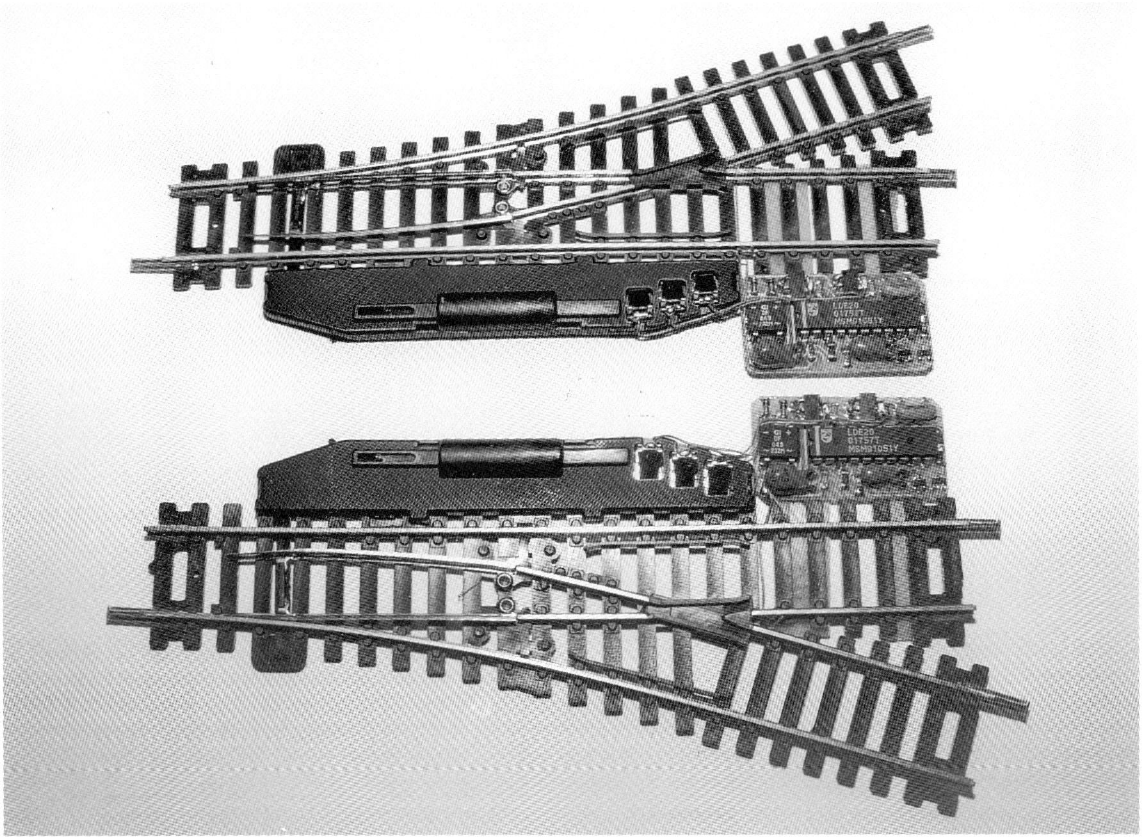

NMRAF 3 is designed for switches with magnetic control. If you use switch motors on your layout, use a module with a higher type-number, such as NMRAF 4. This module is relatively easy to build with surface components and there is plenty of space on the board in comparison to other circuit boards in the series.

group addresses), before returning to the normal service state.

A FEW WORDS ON TROUBLESHOOTING

The risk for faulty soldering work is much less on this module than for the NMRAF 2. In the previous section, I went into great detail on troubleshooting procedure and how faults are remedied. Most of those instructions also apply here.

Because there is no ID check circuitry to base troubleshooting on, lamps are attached to the operating outputs if an oscilloscope does not suffice.

The micro-processor is the same as on the NMRAF 2 but is housed in a smaller casing and has more connecting pins. The program for LDE 20 is also slightly different than for the locomotive decoder's LDE 33.

For correct pin references for the 20-pin IC, see the component descriptions at the end of the book.

SELECTION OF ALTERNATIVE SUB-ADDRESSES

IC 1 has output pins for all eight functions required for four switches. As standard, I have selected switch 1 in the configuration of the NMRAF 3. This implies that pins 15 and 16 are connected to their respective power transistor. If switch 2 is instead chosen as the sub-address, pins 17 and 18 are connected. This means that the lead between Z and pin 15 is cut, and a lead run from Z to L instead. The other transistor is controlled in as similar manner. Here, the lead between Y and K is clipped and Y is connected to M.

To select switch 3, choose pins 5 and 6 and connection points A and B. Finally, for switch 4 (as you may recall, there is space for four switches within the same group), use pins 7 and 8 and points C and D.

WHY ZERO RESISTANCE?

At several locations on the modules, zero resistance is specified. Yes, zero resistance exists. It can be used as bridges at intersecting leads and is also valuable for changing resistance values.

Transistors are normally controlled at their bases via current-limiting resistors. Values vary but are never close to zero. In this circuit, a modern power transistor is employed that requires no resistance. But if you use a traditional transistor (such as a BD 680), it is practical to have space on the board for a base resistor (of approximately 1 KΩ).

When using a BDS 60 transistor, replace the zero resistor with a wire bridge.

NMRAF 3 components list

IC 1	micro processor LDE 20 from Lenz
C1	cap 4.7 µF / 10V tantal
C2	cap 6.8 µF / 35V

IC 1	micro processor LDE 20 from Lenz
C1	cap 4.7 µF / 10V tantal
C2	cap 6.8 µF / 35V tantal
D1	rectifier bridge B50C1500
D2	zener diode 5.2V SOT 23 SMD
D3-4	diode 1N4148 SMD
R1-2	resistor SMD 0Ω (noll)
R3,R6	resistor SMD 100kΩ
R4	resistor SMD 100Ω
R5	resistor SMD 10kΩ
R7	resistor SMD 47kΩ
T1	transistor BC 857
T2-3	transistor BDS 60
X1	crystal 4.19 MHz

Connection points

1	connection to rail
2	connection to rail
3	common output
4	red output
5	green output
6	programming pin
7	programming pin
L,M	connect to Y resp. Z for turnout No. 2
A,B	connect to Y resp. Z for turnout No. 3
C,D	connect to Y resp. Z for turnout No. 4

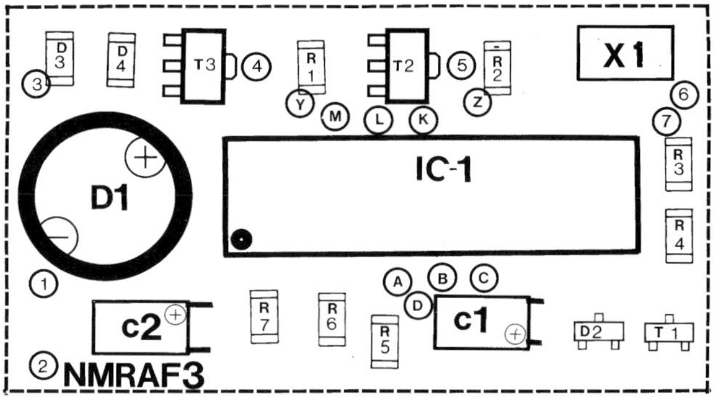

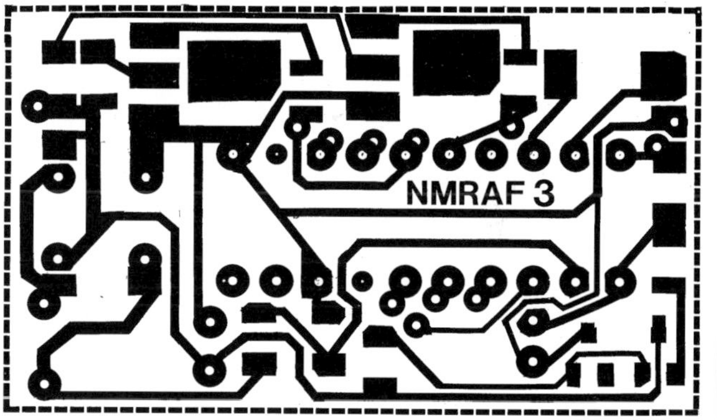

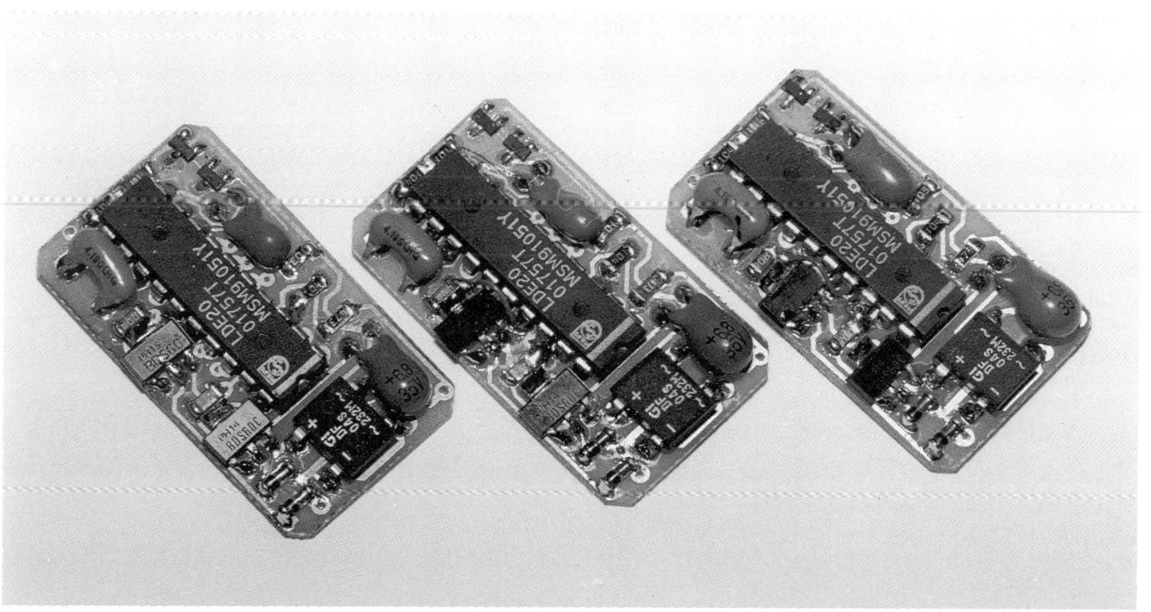

The NMRAF 3 module can perform a number of tasks on the layout and should be mass-produced – mass-production is easy. The main text provides tips on customizing the basic version.

NMRAF 4
Decoder for Four Switch Motors

Both electro-magnetic coils and motors are used to operate model railroad switches. The electro-magnetic variant is by far, the most common of the two . These switches have two coils that throw the switch to its respective end-positions. Switching is accompanied by a loud click and is nearly instantaneous.

Motor controlled switches have been used by the more discriminating modelers for decades. While the first devices were home-built, commercial units are now readily available. Until recently, these were only popular in the US. This method is now gaining acceptance in Europe too.

WHY USE MOTOR CONTROLLED SWITCHES?

The first thing that comes to mind is slow, smooth and realistic switch action. Switch control is quiet and the entire mechanism is hidden from sight, under the baseboards.

Another characteristic is that the motor is stalled intentionally at the end-positions, that is to say, it is run until it stops, with current supply uninterrupted.

Now you are probably saying – as I also said when I first noticed this operating characteristic – you can't do that with an electric motor. And imagine the situation with several switches. It sounds idiotic that one would consciously overheat and burn-out motors.

The finesse is that the motor, with current limitation at approximately 15 mA, holds the switch points against the running rails. Consequently; current supply is secured to the switch points and the risks for derailments caused by wheels "splitting" the switch are greatly reduced.

Supporters of switch motors say that trains never derail. On the minus side, you can probably reckon with increased costs per switch and that you can't force the switch by running through the switch from the wrong track.

The two most well-known manufactures of switch motors are Switchmaster and Tortoise, both from the US. The familiar, green Tortoise profile can be seen in my figures.

THIS REQUIRES SPECIAL DECODERS

You probably understand that one can't use just any decoder to control a switch motor. There are two reasons for this. The decoder must supply a continual voltage of 12 volts to the motor (Tortoise motors are sensitive to improper voltages). Furthermore; when switching, polarity is changed on the *two* motor feed leads as opposed to the *three* leads of an electro-magnetically controlled switch .As of this original writing (April 1994), no switch decoder filling these basic requirements was available commercially. At the time of the translation (April 1997), a few "top-of-the-line-decoders" are equipped with programmable registers. These allow users to choose between electromagnet and switch motor; but at a substancial extra cost.

It took a bit of thought to solve this problem and the first result was the NMRAF 4, a decoder with the service unit constructed entirely with discrete components, that is to say, components such as resistors, diodes and transistors.

REGULATE SWITCHES WITH THE CONTROLLER

I have chosen to control the decoder as a locomotive – a locomotive address is assigned for each group of four

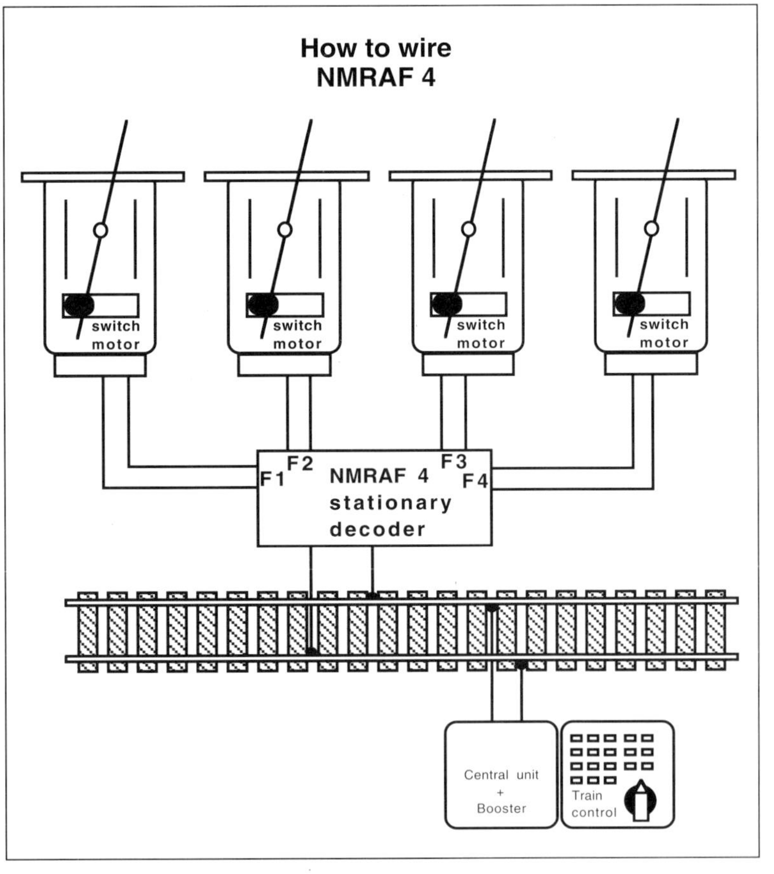

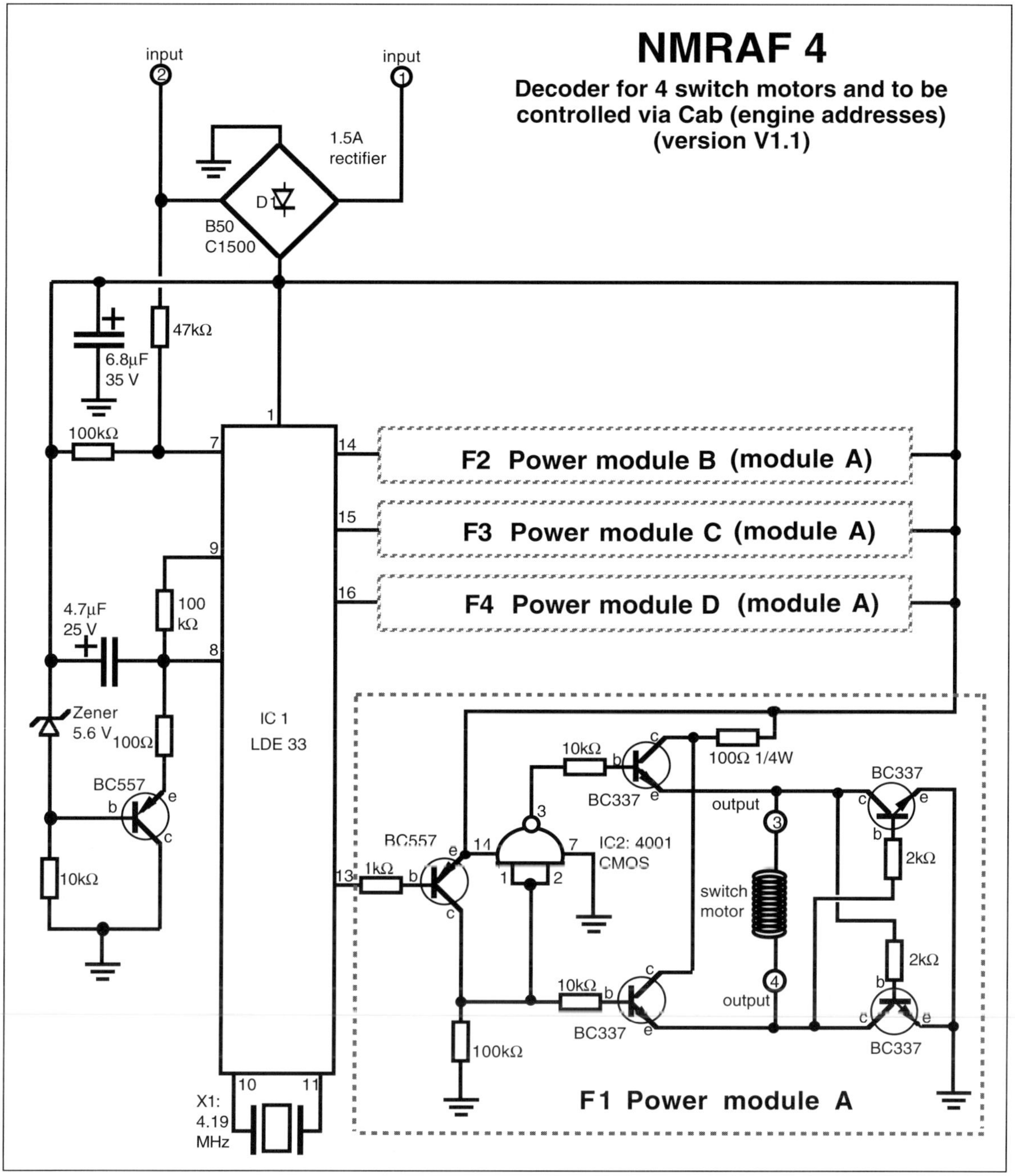

switches and regulated with the function keys on the control unit. This time, you won't have to double-press the keys (as with the NMRAF 2). Each key provides a new switch position and because all function keys have individual indication, it is easy to observe switch status.

One of the four function outputs on the LDE 33 is Pin 13 (F1). On the accompanying schematic, I have presented the four servo units in detail. IC 1 has four gates, meaning that each unit uses just one-quarter of the IC. Additionally, I have used five transistors and a few resistors.

Note that R12 is only required if the switch motor's working voltage is 12 V. For motors needing 16–20 V, the resistor can be replaced with a wire bridge.

AN UNUSUAL BISTABLE SWITCH

After start-up, the function outputs are always high. In digital technology, the basic assumption for all points in the construction is that logic outputs can have either a high or

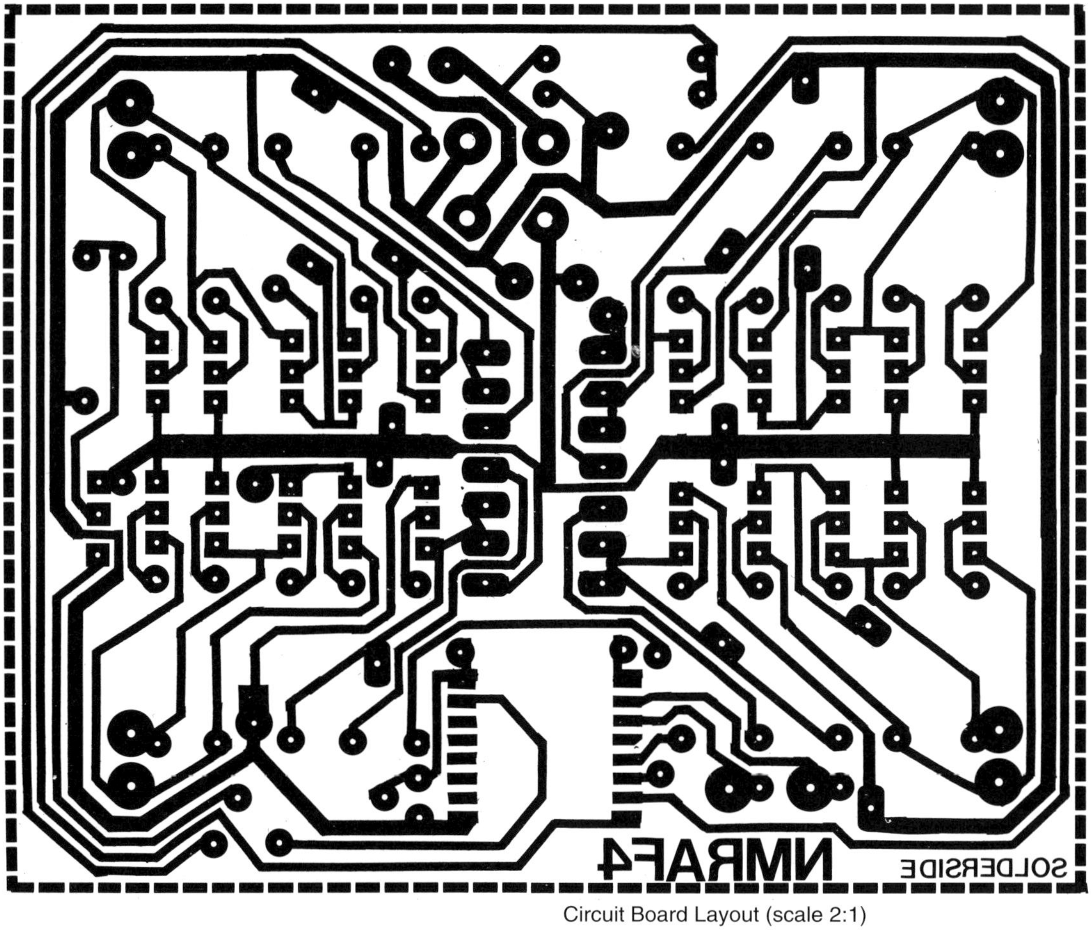

Circuit Board Layout (scale 2:1)

low potential. Low potential is usually the system ground. This can also be referred to as "zero". A point that has a "high" (or "one") logic state usually corresponds to the power voltage's potential.

At Pin 13 is T1, with its PNP function. Initially, it blocks current flow. If T1 had been of the NPN type, it would have passed current between the emitter (e) and the collector (c). Consequently; we also know that the gate (IC 1, functioning here as an inverter) has a low input in that R13 connects to ground. And because the output of an inverter is always the opposite of the input, it is high at start-up. We have thereby defined the conditions for the following bistable switch with four NPN transistors.

The circuit is constructed so that just one of the transistors designated T2 and T3 can be on at the same time – the other is always in the rest state. Furthermore, the same applies for the transistor pair T4 and T5. Only one of the two can be on at a time. Which one, is determined by the state of the T2/T3 pair.

At start-up, the inverter's input is low. Because it is connected to T3's base by way of R3, T3's base is also low – T3 is now off. But its twin T2 is on because the inverter's output is high and it is connected to T2.

Switch Motors – Buyers Guide

Tortoise Circuitron Inc. Dept. 53 Box 322 Riverside. IL 60546 USA	Switchmaster P.O. Box 441432 Aurora, Colorado 800044 USA

AN ESPECIALLY GOOD FEATURE IS THE SMART GROUND CONNECTION

As is clearly shown on the schematic, T2 connects the 12 V feed via R12 to the motor output 3. This is good. But how should we deal with the ground connection to the other motor output, 4? Naturally, it is a prerequisite for the motor to run.

We follow the path of the "12 volt" to T5 through the R5 resistor. T5 is also on now and because its emitter (read up on how transistors work if this is getting confusing) is connected to ground, the collector drops the potential to close to ground (there is an internal resistance in the transistor that generates a voltage drop of about 0.7 V). Consequently; the motor turns in the desired direction until it reaches its "stalling" position, as mentioned at the begin-

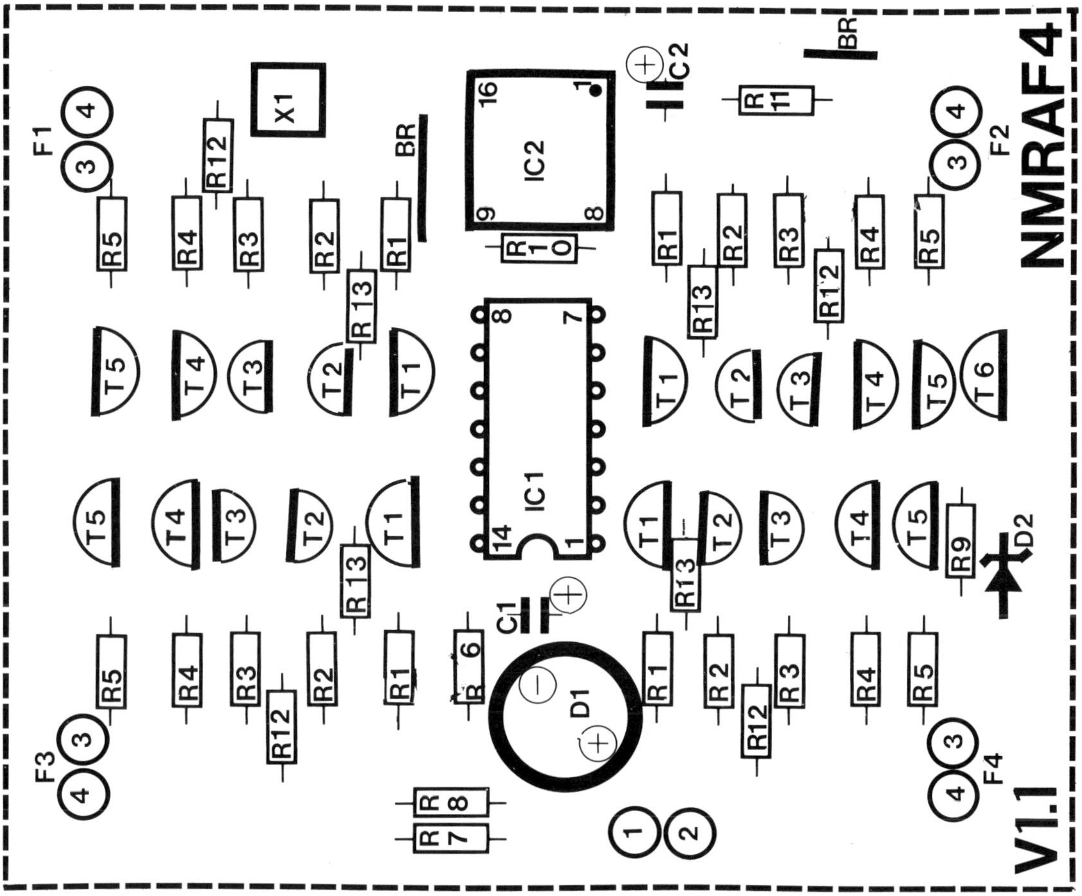

Circuit Board Layout (Scale 2:1)

Components List for NMRAF 4 (Version 1.1)

Components for one module (A=B=C=D)

T1	BC557 PNP-transistor
T2—T5	BC337 NPN-transistor
R1	1 kΩ 1/4 W resistor
R2=R3	10 kΩ 1/4 W resistor
R4=R5	2 kΩ 1/4 W resistor
R12*	100 Ω 1/4 W resistor
R13	100 kΩ 1/8 W resistor
(1/4 IC1	CMOS 4001)
R7	47kΩ 1/8W resistor
R8=R10	100kΩ 1/8W resistor
R9	10kΩ 1/8W resistor
R11	100Ω 1/8W resistor
C1	6.8μF / 35V tantal cap
C2	4.7μF / 25V tantal cap
D1	1.5 A / 40V diode bridge
D2	5.6 V / 0,4W zenerdiod
BR	wire jumper (2 pieces)
*	see text

Common Components

IC1	CMOS 4001
IC2	LDE 33 Lenz CPU
T6	BC 557 PNP-transistor
R6	100Ω 1/8W resistor

Connections

1-2	input
3-4	output to switch motor

ning of this section.

LET'S THROW THE SWITCH

Pressing the control unit's F1 key should throw the switch. What is happening in the decoder? The most obvious is that the output for F1 (our Pin 13) goes "low". Consequently; T1 is suddenly switched on and its collector voltage rises to approximately 12 V. Now the inverter's input is high and its output low. T2 turns off and its duties are assumed by T3 instead. This in turn means that the "old 12 volts", if I may express myself so, to the motor output 3 and to T5's base vanishes. T5 no longer conducts current to ground.

Instead, a "new 12 volts" comes to the motor output 4, in that T3 is now on. The result is that T4's base goes high, T4 leads the current to ground, and motor output 3 becomes a minus conductor.

MINITIP

Control of 3-way switches with NMRAF 3

The NMRAF 3 module has a larger and easier to mount IC than its closest cousin, NMRAF 2. Operationally, the modules are also different. NMRAF 3 is controlled with the keypad in the traditional manner for switches and other electrical accessories.

This decoder is the best choice for those of you with larger layouts or if you have already obtained the necessary equipment.

This minitip shows how to easily wire a 3-way switch using two external components. For more general wiring information, see the minitip that accompanies NMRAF 2.

Programming procedures differ considerably between NMRAF 2 and 3. For NMRAF 3, the procedure involves shorting pins 6 and 7 on the circuit board. For detailed information, read the section covering NMRAF 3.

How to wire a 3-way turnout switch

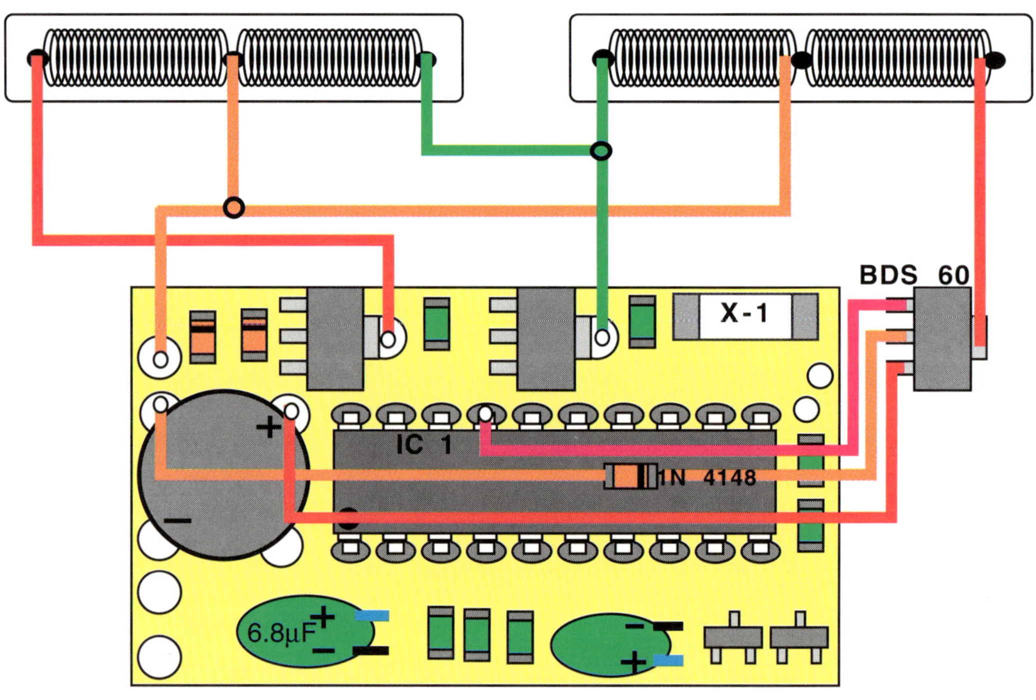

Now the polarities at Points 3 and 4 are switched and the motor can run to the switch's other "stall" position. From here, the switch can now be thrown in the opposite direction. I hope that you have been able to follow me on this electronic tour. If nothing else, it should be of help if troubleshooting is needed.

A FEW PRACTICAL TIPS

The decoder is rather simple to build and has but one surface mounted component. There is also plenty of space to solder IC 2 in place in that it is the only component mounted on the solder side. Furthermore; only half of the pins need be soldered because the others are not needed for this application. Pin 7 is a good place to start soldering.

All other components are hole components and are mounted on the component side. Start with IC 1 followed by the low components (resistors and diodes). Finish up with the transistors, tantalum capacitors and rectifiers, making sure that they are positioned properly – use the accompanying component layout.

Reprogramming of "locomotive" addresses was covered in detail in the NMRAF 2 section. Don't forget that even here, correct address changes can be made even though this is not indicated by the programming unit. To keep component totals to a minimum, I have used an H-bridge for the motors. This os the load that normally responds with a very short current pulse over the programming rails.

MINITIP

Using NMRAF 2 for 3-way switches

The NMRAF 2 module is intended for control of one or two remote functions. Obvious applications are uncoupling ramps and 2-way switches.

By adding just two components, the NMRAF 2 decoder can be used for 3-way switches. This is the subject of this minitip.

At the top of the figure are two double electro-magnets. These are standard for 3-way switches. Normally, one is mounted at each side of the switch. They are drawn in parallel in the figure to more clearly illustrate wiring. The green leads symbolize the switch's middle position.

The F1, F2 and F3 keys are used on the keypad, and as for other NMRAF 2 constructions, double-pressing is required. For more information, see the NMRAF 2 section.

The expanded functionality is attained with just two additional components, a 1N4148 diode and a BDS 60 power transistor. There are a number of ways to connect these components. Here are a few suggestions:

If you make your own circuit board, glue the transistor on a 10 mm (about 1/2") extension to the board. The diode can be soldered to the rectifier.

An alternative method is to mount the transistor at the switch. The drawback is that additional wiring is required.

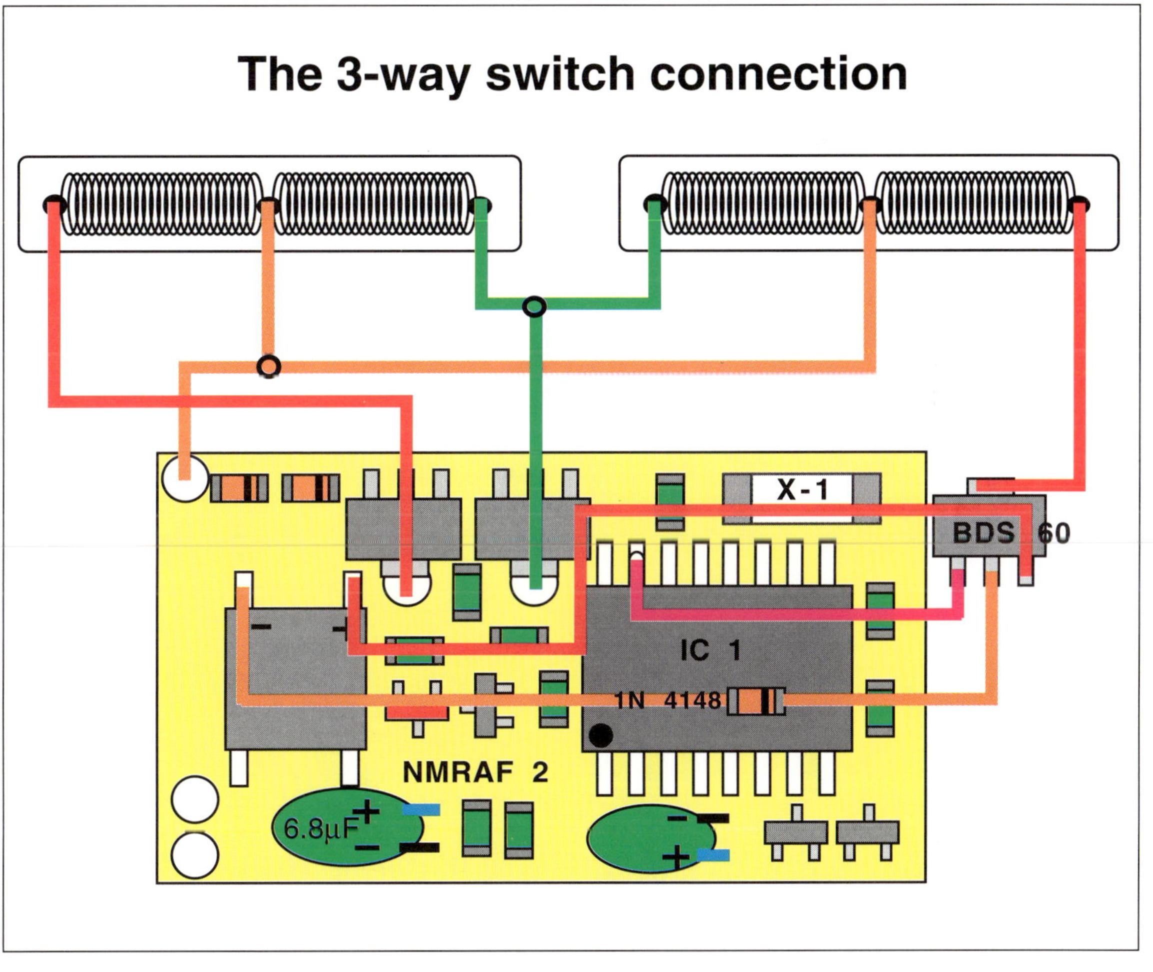

NMRAF 5
Stationary decoder for 1 – 4 switches

Several manufactures offer complete switch decoders for the 2R digital system. These are normally intended for controlling four electro-magnetic switches and are placed under the baseboards at a location selected to minimize wiring.

NMRAF 5 is the decoder in this construction series that resembles these commercial decoders most. They are operated and programmed similarly – operations is via the keypad and programming via an "onboard" programming button. For more information on programming, see the section on NMRAF 3.

FLEXIBLE SELECTION OF TOTAL SWITCHES

One finesse of the NMRAF 5 is that you can customize it as required. You use just the components required for your switch type and number at a specific location on the layout.

One of the benefits of digitizing the layout should be that wiring is minimized. In keeping with this concept, two isolated switches can each have their own decoder. Using all four of the decoders' outputs at all costs, defeats the purpose of the system somewhat by necessitating excess

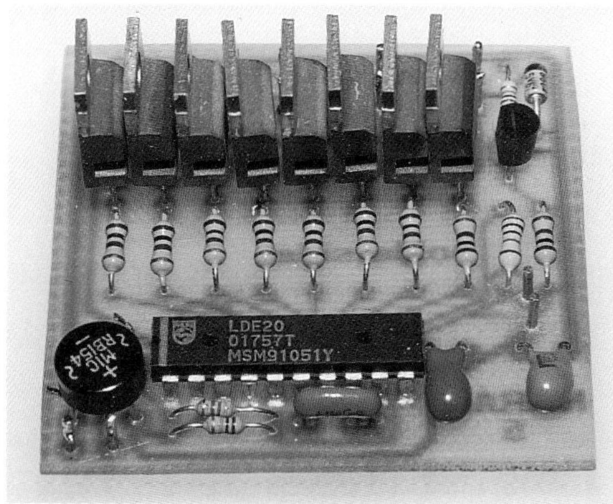

NMRAF 5 makes maximum use of the LDE 20 chip. The module's functionality resembles that of the NMRAF 3 and it is programmed in the same manner. Because four turnouts share a single LDE 20 chip, the module is economical. It is also easy to build. It is constructed using the larger hole components only, which are easier to mount and solder.

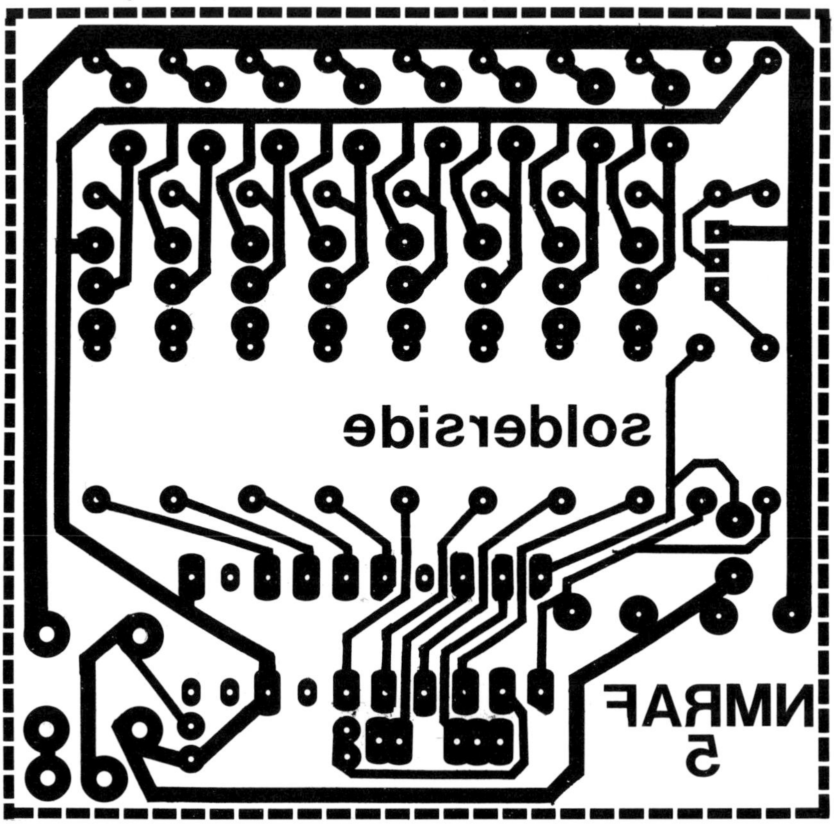

Circuit Board Layout (Scale 2:1)

wiring.

Now you might say that you can't waste switch addresses. You have a large layout and would like to use all 256 addresses.

Fine, I have the solution here in my back pocket. Build two identical decoders. Install them at their respective locations and assign the same group address to both. Using different outputs at each, you are able to use all switch addresses while at the same time, reducing wiring requirements.

Situations can arise where you would like two remote switches to operate in concert with one another – one command throws both switches. This can also be achieved using the this method.

WARNING FOR VOLTAGE DROPS

In HO and larger scales, current is normally led through the rails for booth locomotive and switch control. Only if distances are long or if you are building an outdoor layout, will you need separate drive current supply via heavy feed cables.

The reason for this warning is that with time, each rail-joint becomes a potential fault source. Voltage drops over rail-joints can easily reach 100 mV. If distances are long, the sum of all voltage drops can be considerable.

I once checked the voltage at the furthest reaches on an N scale layout with a booster output voltage of 19 volts – it was 8 volts! The remedy was a two-wire lead around the layout with an area of 4 mm^2 (0.01 sq. Inches), and with

Component List for NMRAF 5 version 1.1

Components, per added switch (1+2 = 3+4 = 5+6 = 7+8)
(Max 4x below)

T2=T3	BDX 34 B PNP power transistor
R1=R2	10 kΩ 1/4 W resistor
D3=D4	1N4148 diode

Common Components

IC1	LDE 20 DIL from Lenz (20-pin IC)
T1	BC 557 PNP-transistor
R3=R6	100 kΩ 1/8W resistor
R4	100Ω 1/8W resistor
R5	10kΩ 1/8W resistor
R7	47kΩ 1/8W resistor
C1	4.7µF / 25V tantal cap
C2	6.8µF / 35V tantal cap
D1	1.5 A / 40V rectifier bridge
D2	5.6 V / 0,4W Zener diode

Connection points

1	power input (digital DCC)
2	power input (DCC)
3	common output for all switches
4	red switch output
5	green switch output
6	programming pin
7	programming pin

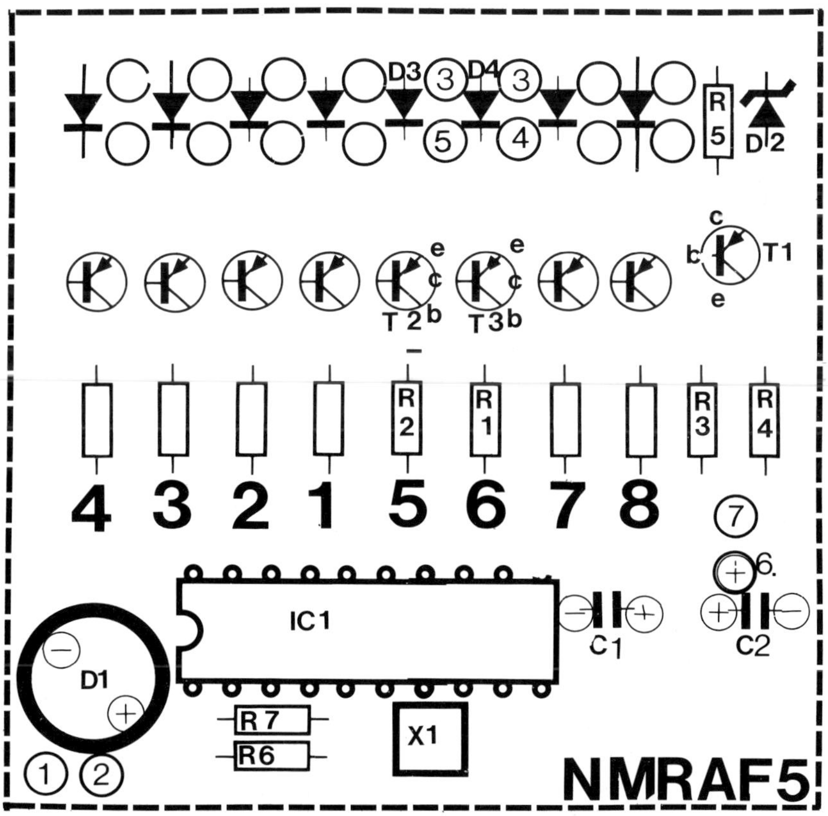

Circuit Board Layout (scale 2:1)

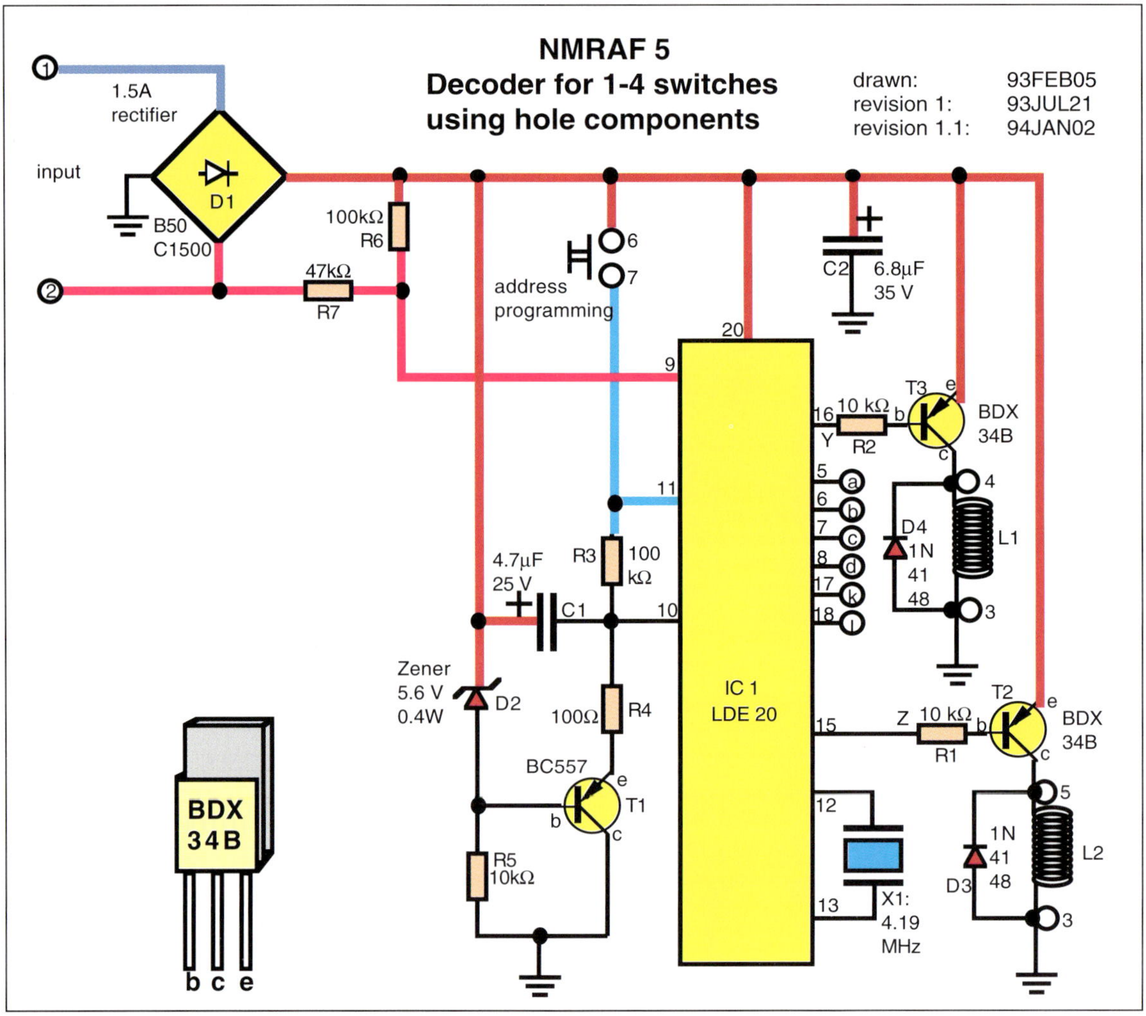

NMRAF modules 4 and 5 have thru-hole components to be soldered on the "backside". The proffesional soldering is reached if each pin before solder is cut to about 0.7mm above surface. However fastest result is achieved if pins/wires are cut after solder. Here on module 4, also the solder work with the SMD-IC is added.

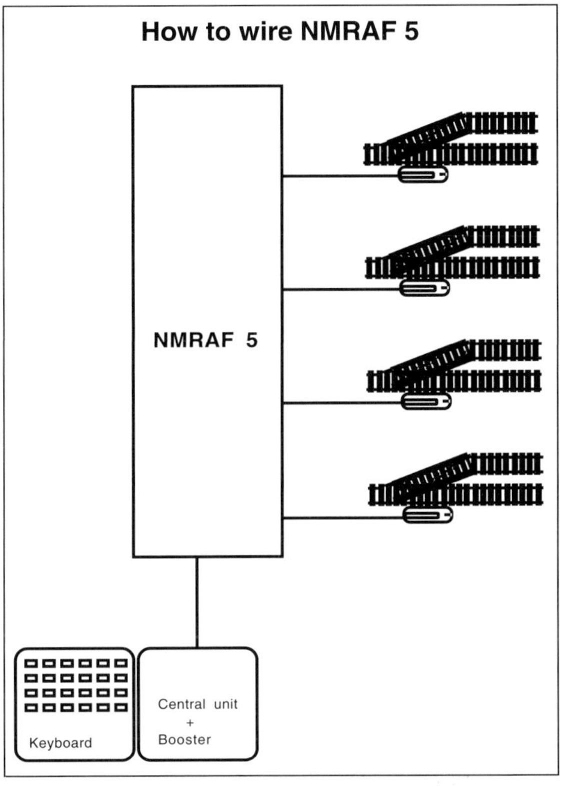

local feeds to the track every 30th rail-joint and to each accessory decoder.

Considering that each output stage in the digital supply system (central units and boosters) can deliver a maximum of 3 A, larger layouts require several, seperate supply blocks. A wise choice would be to feed all stationary decoders from a sepearate booster.

A SUITABLE CONSTRUCTION FOR BEGINNERS

Among all 2R decoders covered in this book, NMRAF 5 is especially well-suited for beginners. It is large, single-sided circuit board, with plenty of working space and is built using hole-components only. And all components of the same type are arranged in rows and facing in the same direction.

The electrolytics can be of the traditional type or tantalium, depending upon what is available. Note that their plus poles face each other. This is also true of the zener diode. It faces in the opposite direction of the other diodes.

A special application could be a 3-way switch that can be easily equipped with the NMRAF 5. In this case, use three transistors in a row to obtain three adjoning key functions.

NMRAF 5 is very flexible and you just mount the amount of power transistors as needed in that "working area". It might be just one 3-way switch and in that case 3 power transistors are used.

NMRAF 6
Decoder for two switch motors

In the NMRAF 4 section, I described in detail the differences between electro-magnetic and motor controlled switches. Here is an additional decoder for switch motor control. The following differences motivated design of the NMRAF 6:

NMRAF 6 is controlled via the keypad and not the control unit.

NMRAF 6 is constructed of hole components only (easier to build)

NMRAF is constructed with fewer components, thanks to the operational amplifier.

NMRAF 6 is a smaller module for 1–2 switch motors (3-way switches, for example).

Moreover, there is a completely different philosophy behind this module, with more integrated functions and fewer components.

OPERATIONAL AMPLIFIERS CAN HANDLE LOADS

Operational amplifiers are often used in measurement and audio applications. It isn't as common to use them to control motors. This is exactly what we will be doing here and, as you will see, the combination is superb.

The IC we will be using is a sheep in wolf's clothing. It can handle large loads despite its small, uncooled casing. The output varies between logic high and low levels and the operational amplifier is wired – via gates **n1** and **n2**, that make up a bistable switch – so that output **n5** is low when **n6** is high, and vice versa.

The prerequisite for the motor to rotate is that the motor connections 3 and 4 are at different voltages. And when these two change polarities, the motor's direction of rotation also changes. As I mentioned earlier, the electrical and mechanical functions for switch motors are covered in detail in the NMRAF 4 section.

SELECTION OF SUB-ADDRESS AND CASING

In its standard configuration, the circuit board is designed for the two first sub-addresses within one and the same group address. This means that we use output pins 15–18 on IC 1. If you would like to change to the two highest sub-addresses (two other switches), a few minor modifications are made on the circuit board's soldering side. A figure is provided that shows the process in detail.

There are several different manufacturers of this popular IC. Some produce the chip with an 8-pin casing, others with a 16-pin casing. Although they both contain two amplifiers, the discrepancy in pin total means that they should require two different circuit board layouts. I've been a bit lazy and made a basic configuration for the 8-pin casing and a variant for the 16-pin version. A figure is provided showing the necessary modifications.

A FEW PRACTICAL TIPS

This is a particularly easy to build and uncomplicated decoder, especially if the 8-pin casing is used and the two lowest sub-addresses are selected. This makes it a straight mounting and soldering operation. Begin construction with the low components, such as resistors and diodes, and continue with the ICs. Lastly, install the bridge rectifier and capacitors.

There are several manufacturers and configurations of crystals. I normally use a resonator (ceramic crystal) from Murata, but this isn't critical. Although some crystals have three connections, it is more common with two. Even the 3-pin variant can be used if the middle pin is connected to the signal ground (pin 10 on IC 1, in this case). In these 3-pin crystal cases are two, very small capacitors.

The connections on the board to the various leads are round soldering tabs with center holes. I usually enlarge these holes with a 1.2 mm drill, insert small connection pins from the board's solder side, and solder them in place. This eases construction later in that you won't need to turn the module each time you need to connect or disconnect a lead.

I usually solder a closing push-button switch to connections 5 and 6. This simplifies re-programming considerably. If you only expect to make address selection once and for all, a short-circuiting, miniature screw driver can provide the same service.

How to wire NMRAF 6

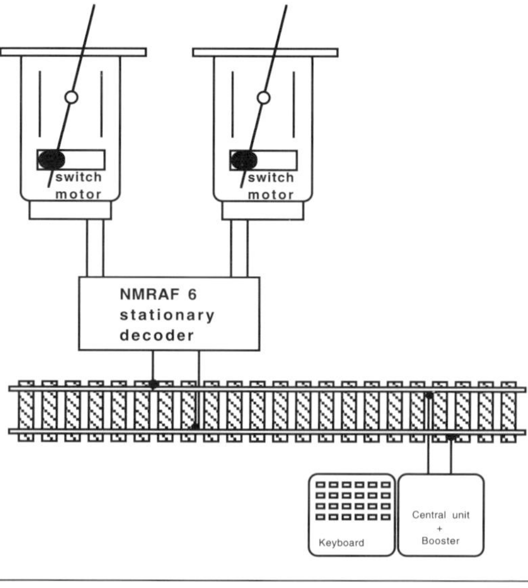

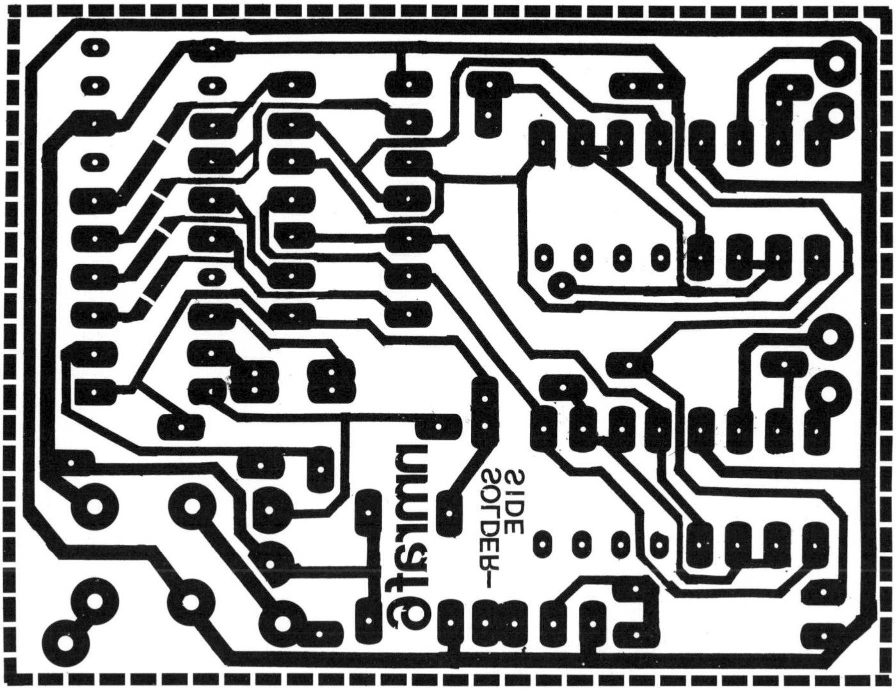

Circuit Board Layout (Scale 2:1)

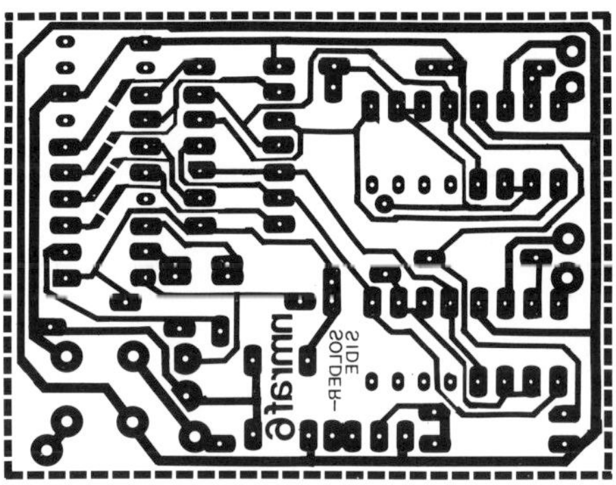

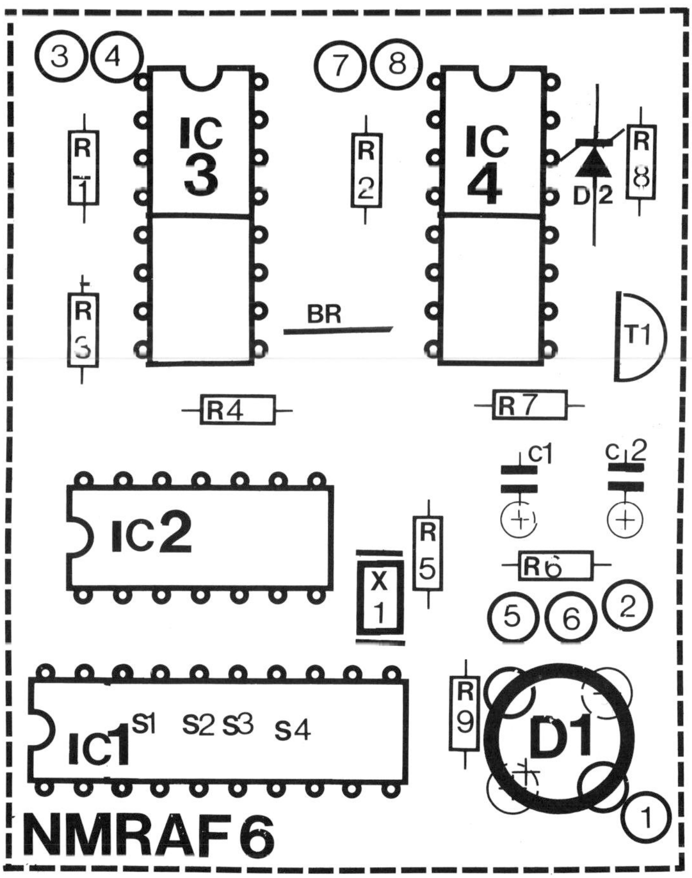

Components list for NMRAF 6

IC1	LDE 20 DIL casing Lenz (20-pin IC)
IC2	CMOS 4011
IC3-4	L 272 dual op amp-IC
T1	BC 557 PNP-transistor
R1=R2	100 Ω 1/4W resistor
R3 -R5	100 kΩ 1/8W resistor
R6	47kΩ 1/8W resistor
R7	100Ω 1/8W resistor
R8	10kΩ 1/8W resistor
R9	100 kΩ 1/8W resistor
C1	4.7µF / 25V tantal cap
C2	6.8µF / 35V tantal cap
D1	1.5 A / 40V diode brigde
D2	5.6 V / 0.4W Zener diode
X1	4.19 MHz X-tal
BR	wire jumper

Connection points

1 and 2	Input
3 and 4	Output 1 for switch motor
5 and 6	Programming pins
7 and 8	Output 2 for switch motor

IC 3 & 4 are available in two variants

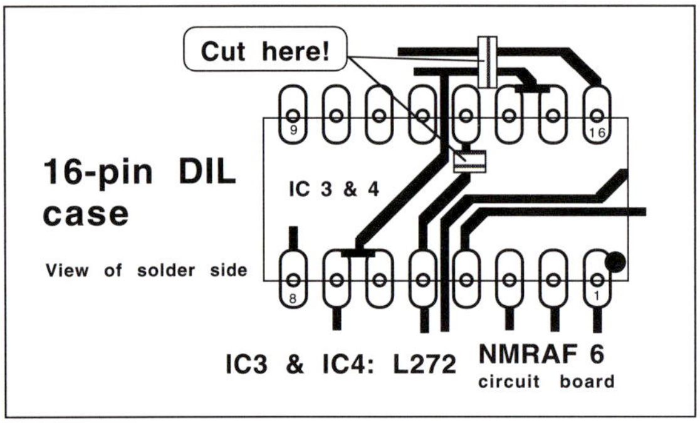

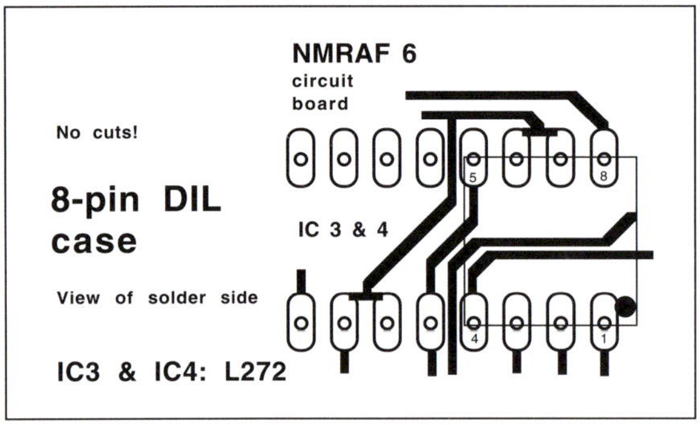

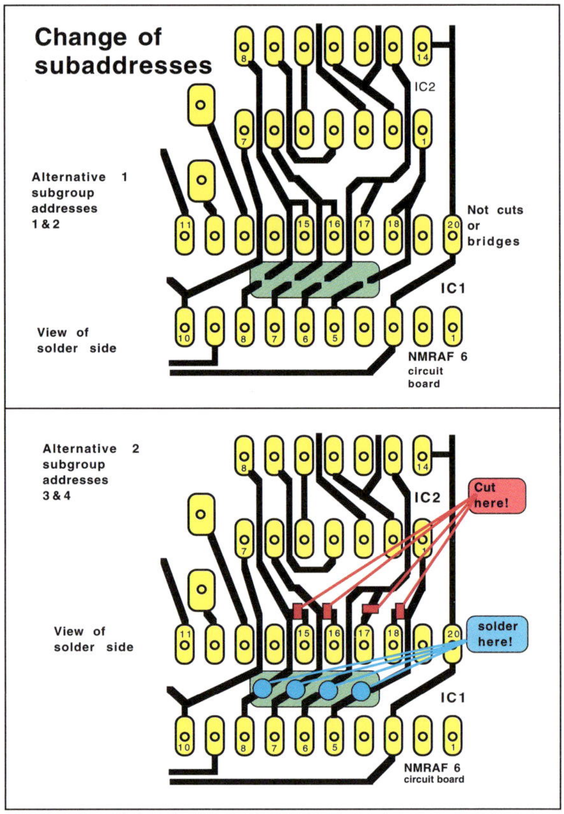

MINITIP

H-bridge in N scale

H-bridge is the collected term for a number of connection variations for regulating electric motors. The circuitry surrounds the motor in an H-form.

Briefly, functionality allows one of the two transistor pairs (T1 and T4, or T2 and T3) to be on at the same time. Current can be fed to the motor's left pole by T1, and to ground by T4 to rotate the motor.

When T2 and T3 are on, the motor rotates in the opposite direction, following the same operational characteristics as for T1 and T4.

The IC LI 11 is a special drive circuit used in several applications. For more information, see the NMRAF 1 section. The IC is connected in turn to LDE 33, the decoder's main IC. Thus, a practical little decoder is attained when NMRAF 1 is combined this H-bridge.

There is an alternative for the diodes that provides compact construction using SMD cojmponents. Check the component descriptions at the end of the book. There is a component, L 21 (the designation on the casing), that contains dual diodes (D1 and D2, alternatively D3 and D4) that will enable you to save space.

H-bridge for N-engines

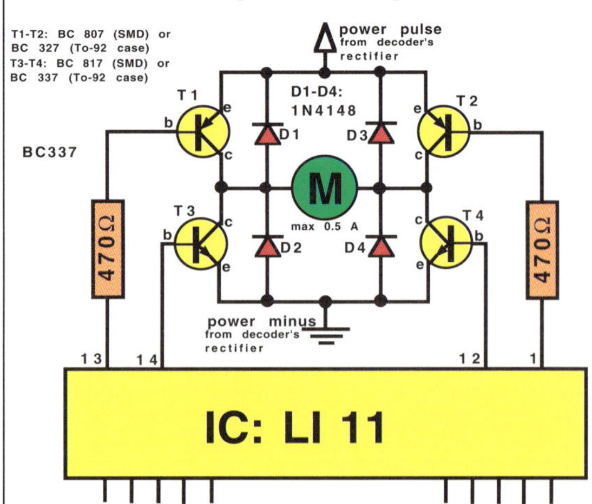

NMRAF 6
Stationary decoder for two switch motors (Tortoise type)

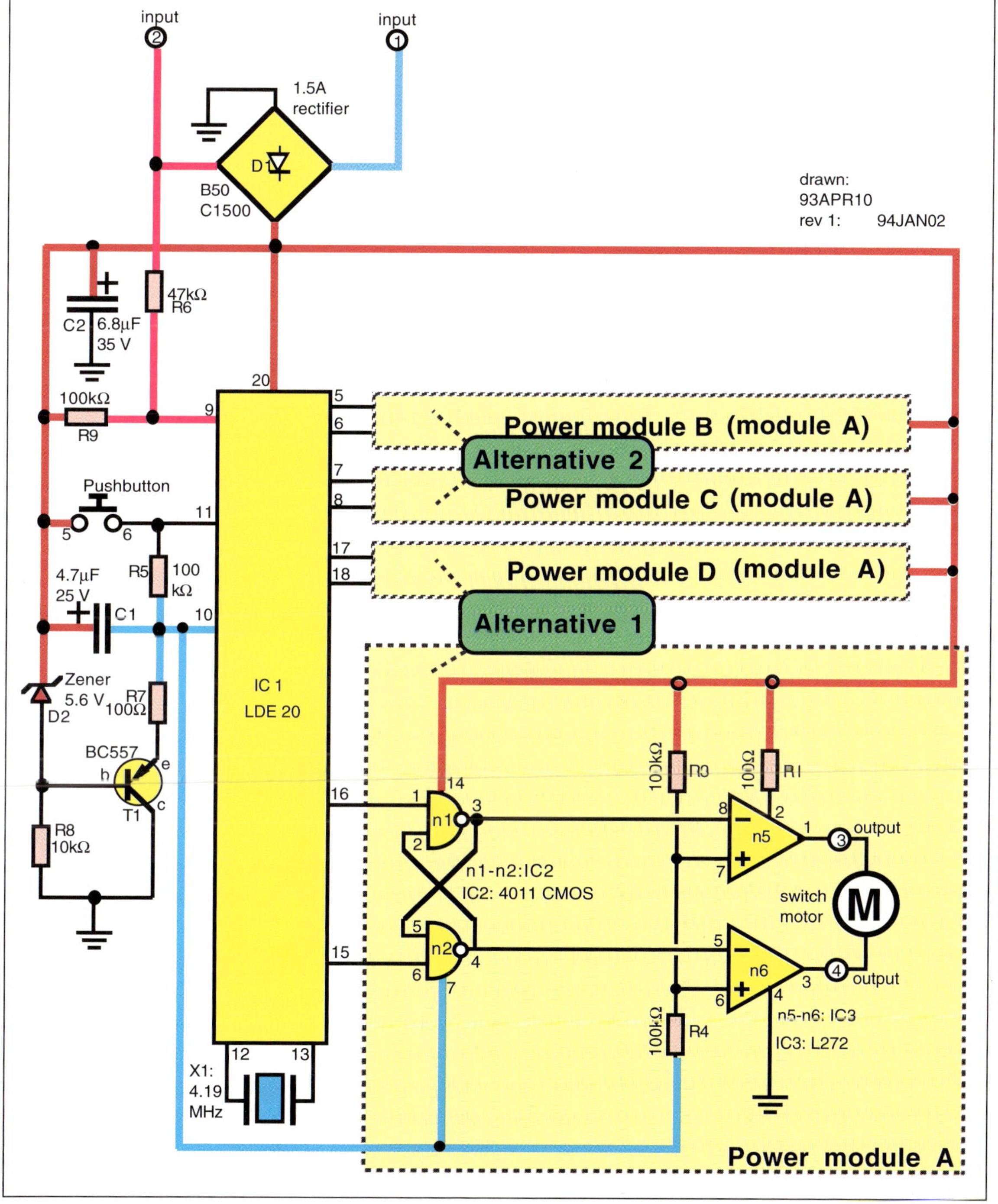

NMRAF 7
Universal booster with 3A output

Actually, this module could be called a power amplifier, but in model railroad digital contexts, the term *booster* is more common. In principle, the module amplifies the weak digital signals from the central unit, from 5 volts and a few milliamps to approximately 17 volts and 3 amps. Power is sufficient for five normal N scale locomotives or three normal HO locomotives.

The booster is simple in construction and does not have all of the features of commercial boosters. There is a big brother to this booster in NMRAF 8 (not in this book), with more capacity and features.

SHORT CIRCUIT PROTECTION

NMRAF 7 is constructed around the L 298 IC, a driver with a full H-bridge that can manage a maximum of 3A. The circuit functions within the 0° to 70° C temperature range and features built-in overheating protection. All the circuit's control inputs work with 5 volt logic (often referred to as TTL, Transistor Transistor Logic) and the power unit can regulate voltages from 5 to 46 volts.

In addition to the built-in protection, the circuit is equipped with a transient protection in the form of the D6 rectifier at the output. It is easiest to drive the module using a power pack supplying at least 12 volts AC. The lighting outputs of a common power pack are perfect. Other approved transformers will also work well as long as the output voltage does not exceed 20–22 volts AC.

After voltage rectification, are two parallel voltage regulators, one for 5 volts and one for approximately 17 volts. In an accompanying figure, voltage conversion is shown in detail.

A third function part is also on the same circuit board, namely that for data communications. An opto switch (IC 3) provides galvanic separation between the in-signal and the rest of the construction. Additionally, there is an indicator with a green LED that lights only when the correct data code type is sent through the input.

ALL ON A SINGLE BOARD

The entire module fits on a single-sided circuit board measuring 64 x 60 mm (about 2 3/8" x 2 1/2"). Although the components are few in number, many are quite large and it is therefore important to mount them in the correct order. It is appropriate to begin with the small, horizontal components such as resistors. After this, ICs 3 and 4 are soldered next to IC 5. Now the other components can be mounted, saving the large electrolyte capacitor and rectifier for last. When all of the components have been soldered in place,

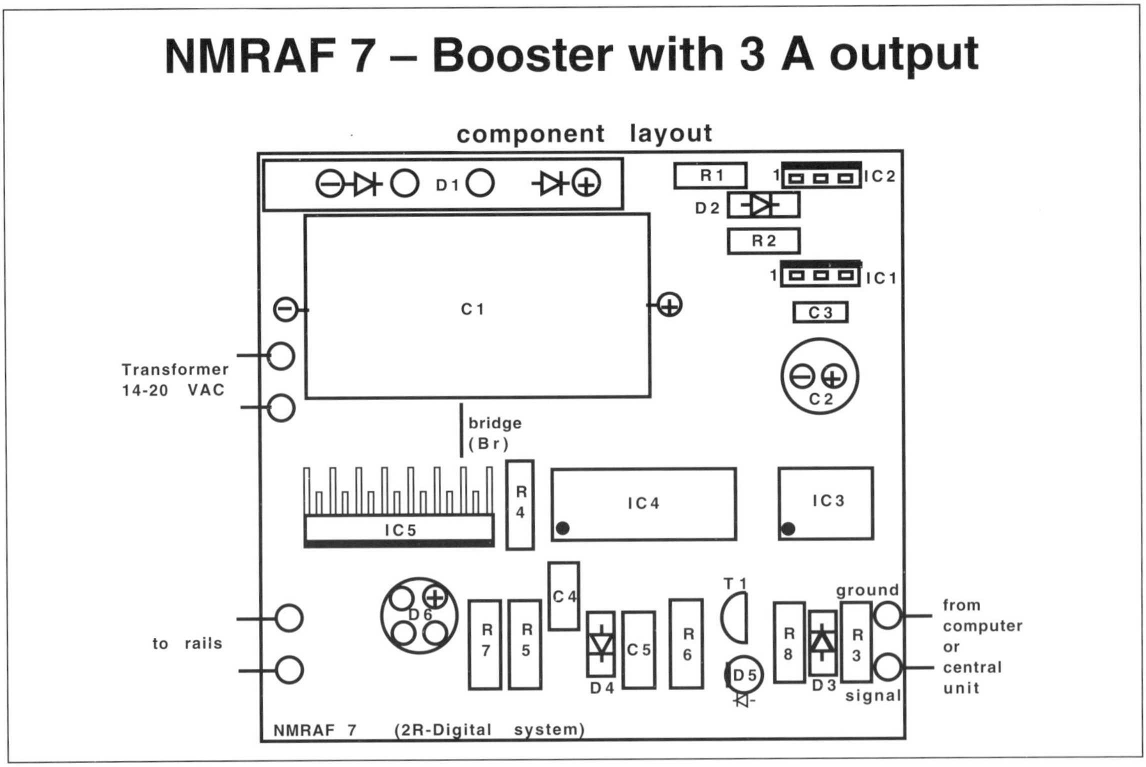

Circuit Board Layout (Scale 2:1)

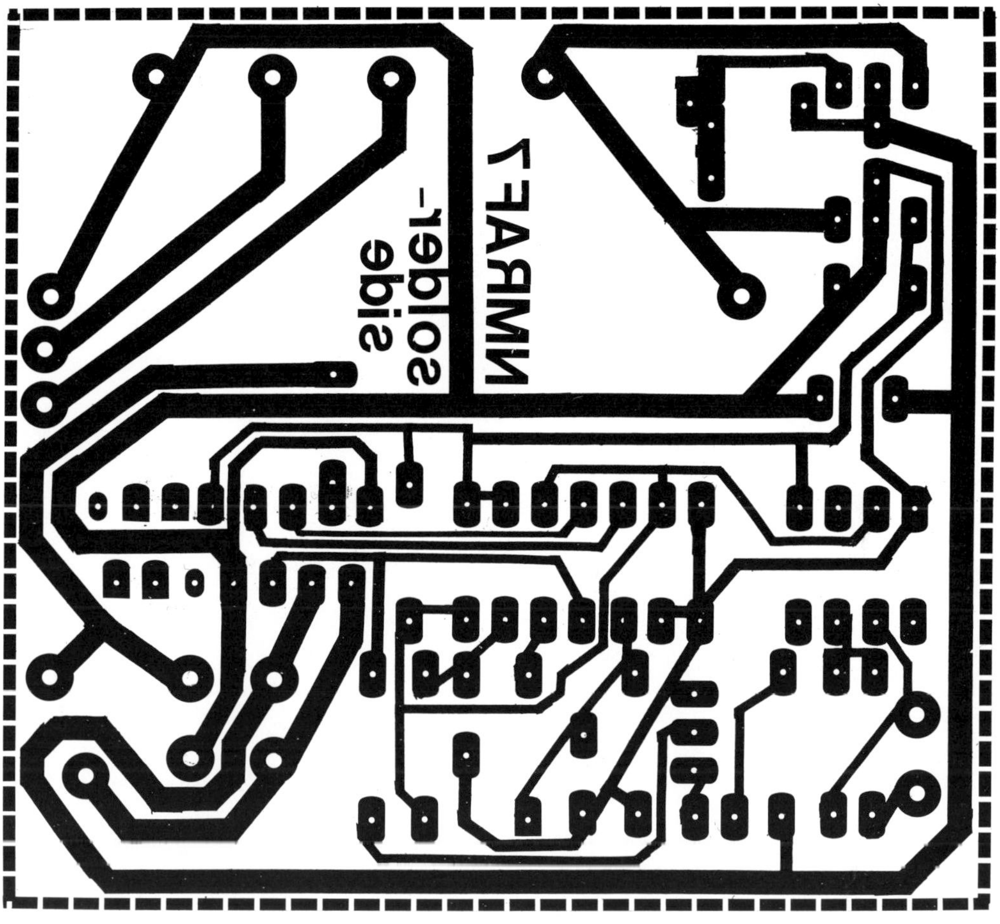

NMRAF 7 - special booster

Components list

IC1	7805 voltage regulator	C3	100nF capacitor
IC2	LM350T voltage regulator (Heat sink!)	C4	10nF capacitor
IC3	6N138 opto-coupler	C5	220nF capacitor
IC4	4070 CMOS Exor-IC	R1	220Ω / 0.25W resistor
IC5	L298 H-bridge Power IC (Heat sink!)	R2	2.7kΩ/ 0.25W resistor
D1	diode bridge 80 V- 4 A	R3	1.5kΩ/ 0.25W resistor
D2	1N4001 diode	R4	180Ω/ 0.25W resistor
D3-D4	1N4148 diode	R5	100kΩ/0.25W resistor
D5	LED, green 5 mm	R6	20kΩ/ 0.25W resistor
D6	diode bridge 100V- 1.5 V	R7-R8	10kΩ/ 0.25W resistor
C1	2200 μF / 40V cap	T1	BC 547 transistor
C2	220μF /25V cap	Br	wire jumper

the two heat sinks can be attached to their ICs.

The module is connected to its central unit with a two-wire cable. The connection closest to the corner on the circuit board is the signal lead and the other control input is ground. Two connections close to the large capacitor are for feed from the power pack. The outputs next to IC 5 are the outputs to the track.

Test functionality using a short length of track and a digital locomotive. The control leads are connected to the booster output on the central unit. Note that there is usually a third connection point, a feedback to the central unit from the booster if an overload occurs that lasts longer than a few seconds.

For more advanced connection alternatives, please see the NMRAF 8 section. This somewhat large and more powerful module is planned for inclusion in the fifth book of this series.

NMRAF 7 Voltage converter

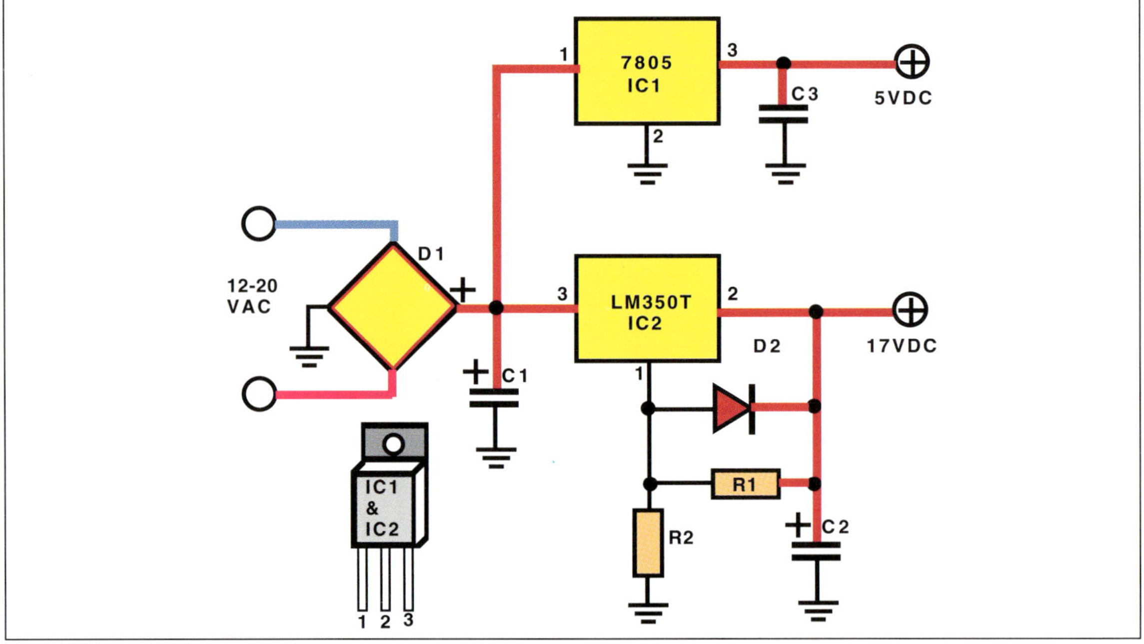

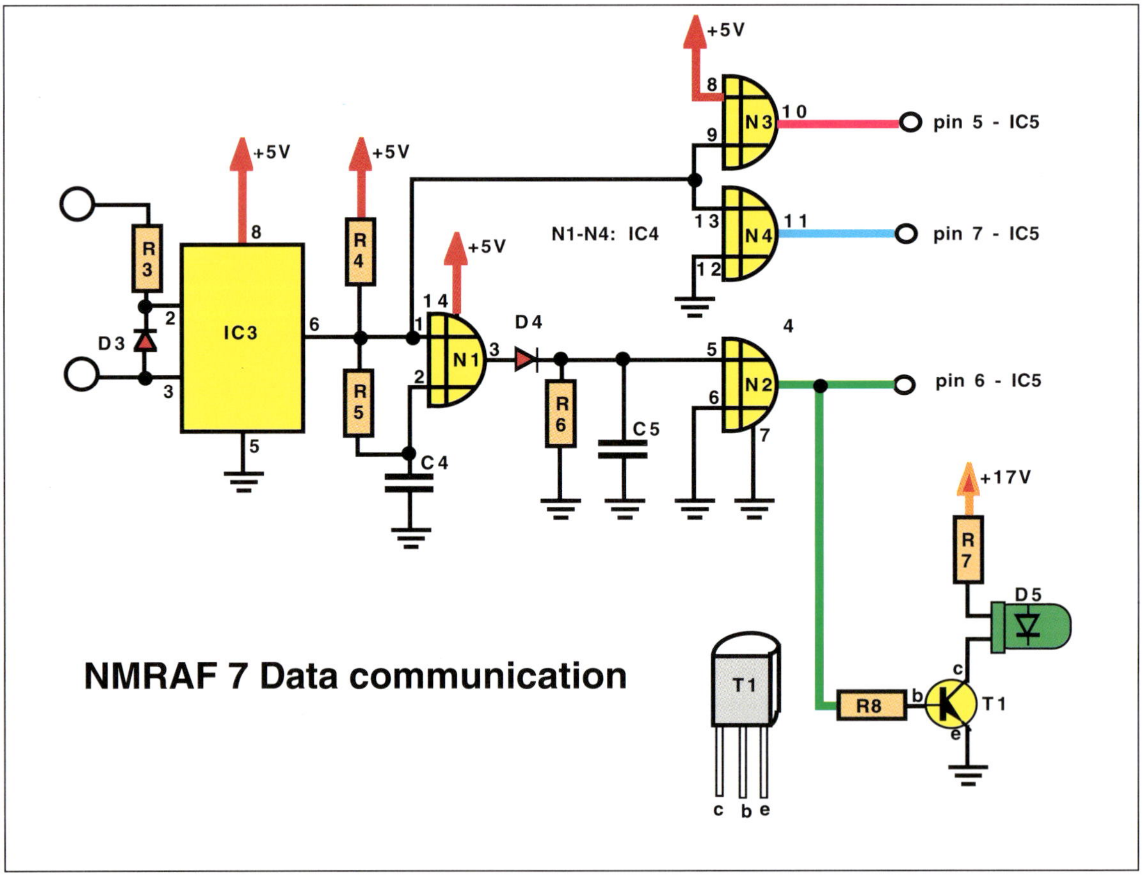

NMRAF 7 Data communication

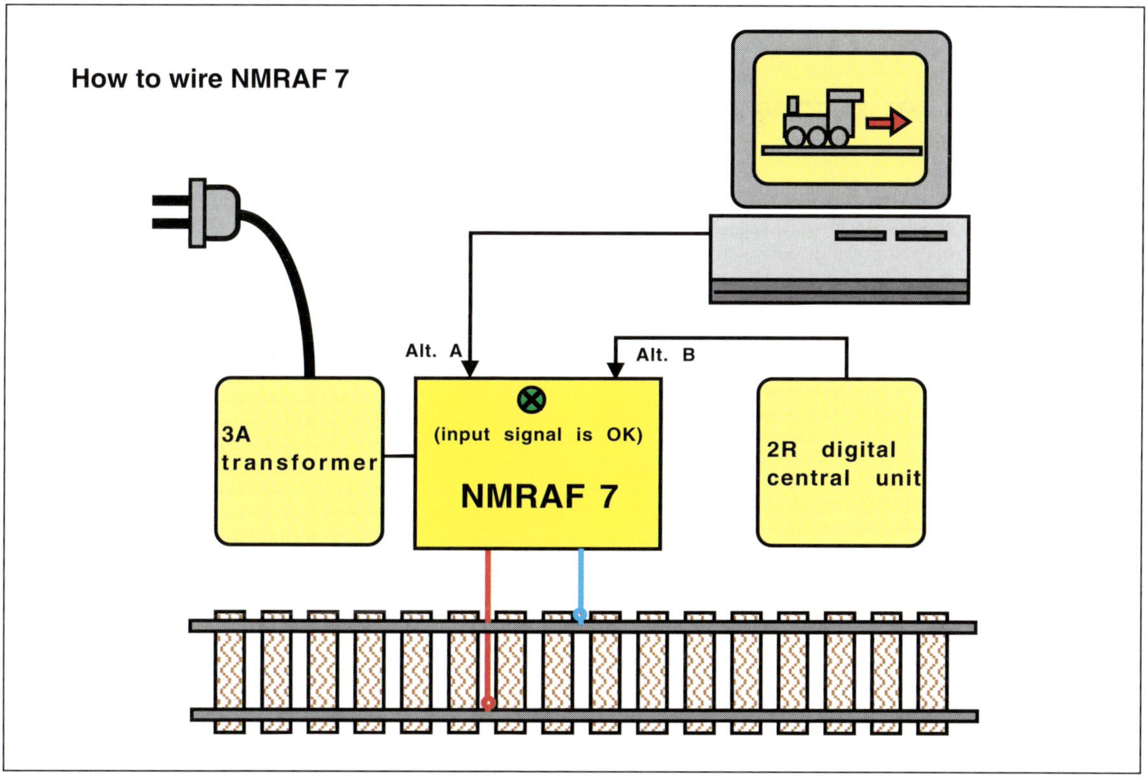

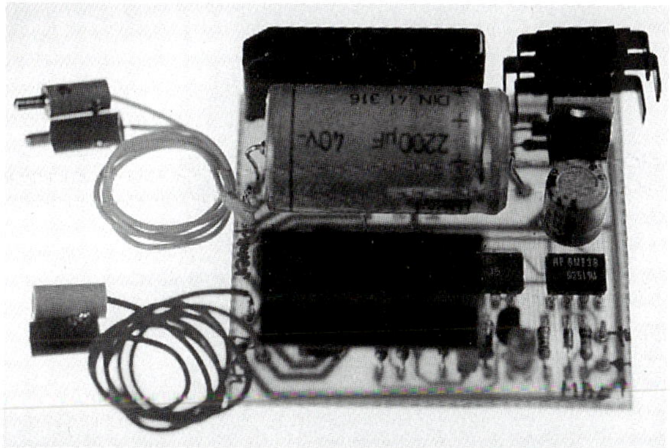

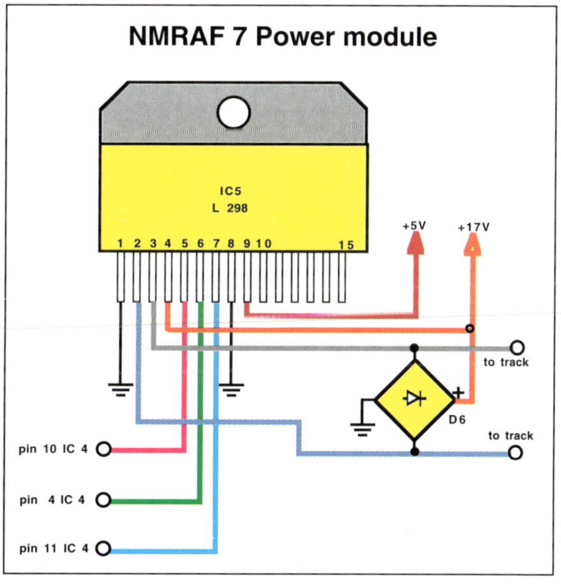

The key components for NMRAF 7. The L298 IC is the power component (to be cooled) for the 3A output and the 3-pin IC LM317 that matches the voltage level to that of the MR-system even if a transformer with somewhat too high output is used.

Building and Operating with 3R-Digital

In the following section, we will cover construction projects for Märklin's 3R (three-rail) digital system (or the *Motorola* format, if you prefer). The basic technology and theory are detailed in the second book of this series. What is provided here is certain additional information for each project.

Märklin introduced their digital control system for HO back in 1984. It is based on special technology developed by Motorola, one of the world's leading electronic companies.

The system consists of software and hardware that communicate with one another via the rails, the bearer of both electrical power and information (read *instructions*). The software is a protocol with 9-bit information, sent out to all decoders from a central unit.

Four of the bits in the pulse train make up the address and use *trinary* logic, i.e., the numeration system based on 3. It can provide a total of 81 addresses (3x3x3x3=81) of which 80 can be used for locomotives. The remaining 5 bits, in this first generation, use *binary* logic and transfer data for commands, etc.

In this book, we will concentrate on projects that often require several units. This is when the most money can be saved by building yourself. Many projects are unique – there are no corresponding units commercially available. At least this was the case when this book was written.

Finally, it should be noted that several projects are constructed using surface components so that the modules will be small enough to mount in HO models. See the earlier section in this book describing the special techniques used with surface component construction.

MINITIP

DC operations with 3R digital systems

Märklin began their 3R era in 1984. That's nearly 15 years ago. At that time the system was polarity-dependent, making it unusable on DC layouts, i.e., systems with current collection at the running rails. Why did a mechanical difference in current collection between 2R and 3R effect utilization of this control system? Because, as you can see by the upper example in the figure, the connection pole to the collection shoe is always connected to a defined pole on the trackwork – the third rail – regardless of the direction in which the locomotive is placed on the rails.

Märklin's system was polarity-dependent and could not handle reconnection of the collection shoe feed to a running rail. Theoretically, you could run a 2-rail engine with this system if you could be sure of always placing it in the same direction on the tracks. For all practical purposes, this excluded the use of turning loops and the like.

A while back, Märklin added more modern electronics to their system, making it possible to change polarity at will. By the time you read this, Märklin will probably have already introduced their 3R system for 2R layouts in the larger scale 1 (a scale that Märklin also manufactures products for).

How do you know if your equipment is polarity dependent?. There are two ways of finding out and we need only direct our attention to the receiver and locomotive decoder. One way is to open the locomotive and identify the decoder's ICs. The 70113 designation signifies the new version.

More modern IC circuits may have come out after this book's publication. The only foolproof method of identifying the system type is to reverse the polarity to the tracks. This is done most easily with a length of track, separated from the layout, and just one test engine on the track. Feed the track with the leads reversed from the central unit – red to brown and vice versa. If the engine still operates, the new technology is used.

This means that DC pikes can be controlled with the 3R system. Even older types of control units can be used, but you must be certain that all decoders are of the newer type.

Two rail engines can use the new Motorola format decoders

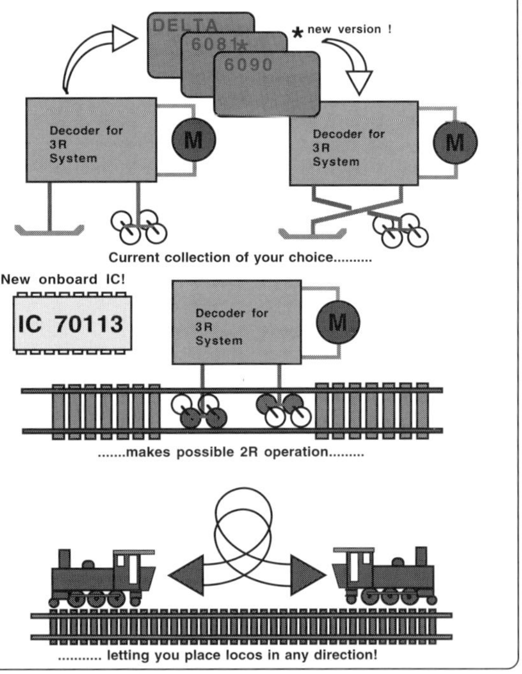

Switch decoder for tight quarters

Thanks to its modest surface area, just 50 x 15 mm (about 2" x 5/8"), this decoder is easy to tuck away beside or under a switch. It can even be placed in half of a track section if you are using Märklin's M-track.

The accompanying photos show how to install the decoder in a Märklin double slipswitch. Even though double slips have little space to spare for add-on electronics (in spite of their relatively large size) there is just enough room for this decoder.

As with the other BVE modules in this book (this module is designated BVE 9009:2), the decoder is coded with solder bridges between various soldering tabs. Main address selection is accomplished with bridges on the small coordinate grid adjoining the feed leads. Each soldering tab in the middle represents one of the 4-position codes (see Märklin's instruction book for the digital system for more information).

In a corresponding manner, selections of sub-groups are coded for straight (green) and diverging switch positions. There are four keys on the keypad for each sub-group.

HOW IT WORKS

The decoder is constructed so that the brown power lead delivers plus voltage to the entire module, including the controlled switch. The red lead supplies 18 minus volts. This is used as the "power zero" for activating the switch's electro-magnetic coils.

The module's "onboard" electronics operate on an unpretentious 8.2 volts supplied via the R1-D1 and C3 circuitry.

IC 1 receives and reads all incoming pulse trains continually and detects when incoming code is in agreement with its own preset code (as set with the solder bridges). When matching code is detected, it immediately reads the relevant command. To carry out the command, it must receive the code twice. It can then assume that transfer was correct and without interference.

In our case, the command means that output pins 12 and 15 on IC 1 are configured in a combination of low and high output levels that IC 2 converts to addressing of a sub-group.

To throw the switch, two commands are sent. The first is an action command. A fraction of a second later, comes a reset command that returns the output to the deactivated state. The electro-magnet needs a short voltage peak to throw the switch. If current is applied for any length of time, the will coil overheat. You'll notice your mistake by the unmistakable odor of burned insulation.

A FEW TIPS BEFORE STARTING CONSTRUCTION

This decoder has components on both sides of the circuit board to make it as compact as possible. From the construction standpoint, this means that means that you can think of the module as having two circuit boards. One is a little more difficult to work with because the surface is not as stable as the other (there are components on the opposite side).

Begin soldering with IC 1. Use pin 14 as a centering point. After IC 1 and before IC 2, mount R5 and C5. Their positions would be otherwise blocked by IC 2. You now solder IC 2 in place, followed by D1 (its position is a bit troublesome), before soldering the rest of the components to the component side. On the solder side, mount T1 first and C1 last.

When the pros make circuit boards, the boards are etched

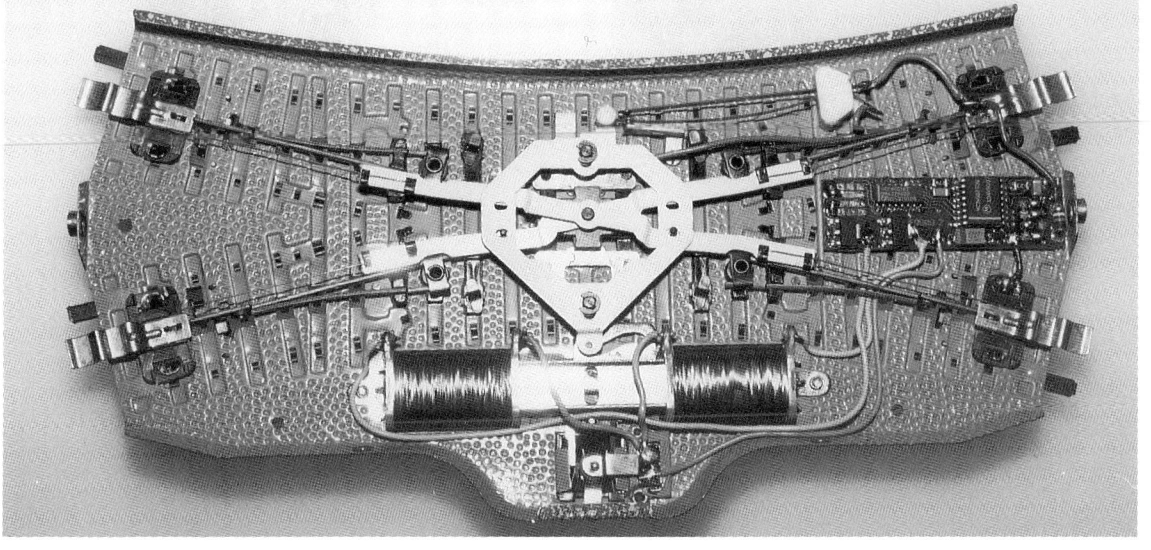

A look underneath the turnout cross. There is really just one position where to place the decoder.

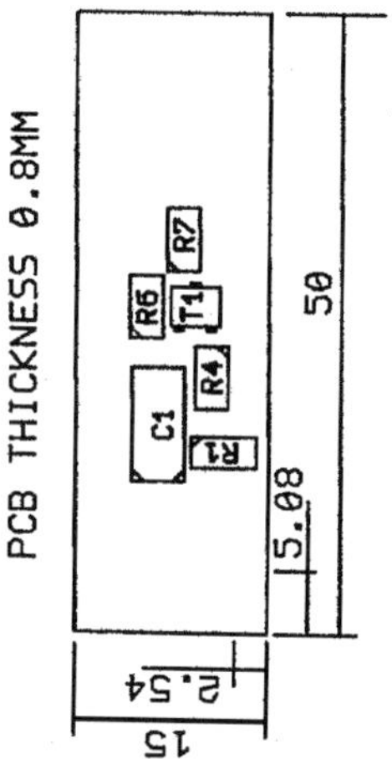

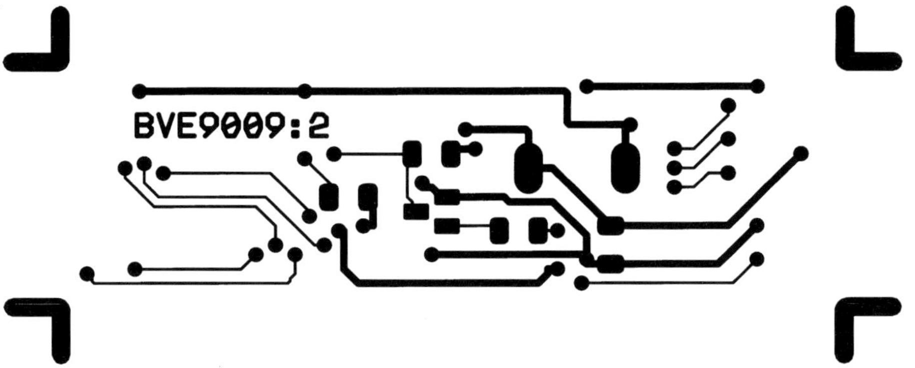

BVE 9009:2
Solder side (2:1)

Switch decoder components list

IC 1	45027 SMD
IC2	4051 SMD
D1	Zener diode C8V2 SMD
D2	power diode BYM10-400
R1	resistor SMD 2.7kΩ
R2	resistor SMD 270kΩ
R3	resistor SMD 12kΩ
R4	resistor SMD 100kΩ
R5	resistor SMD 270kΩ
R6	resistor SMD 100kΩ
R7	resistor SMD 5.6kΩ
C1	electrolyte SMD 15µF/ 25V
C3	electrolyte SMD 10µF/16V
C4	capacitor SMD 3.3nF
C5	capacitor SMD 3.3nF
T1(=Q1)	transistor SMD BC 847
T2(=Q2)	transistor SMD BDS 60
T3(=Q3)	transistor SMD BDS 60
1x	BVE 9009:2 circuit board

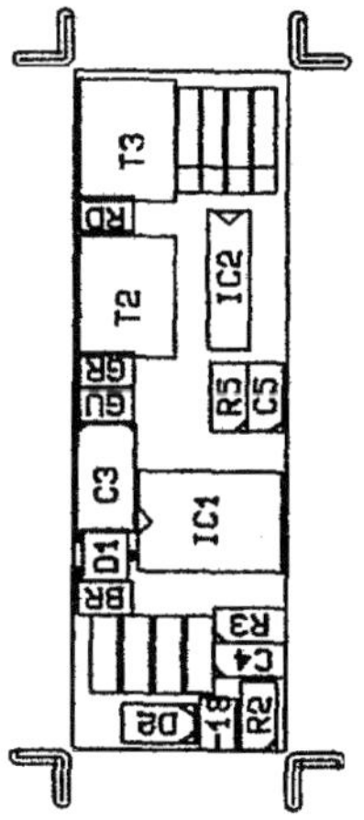

BVE 9009:2
Comp. side (2:1)

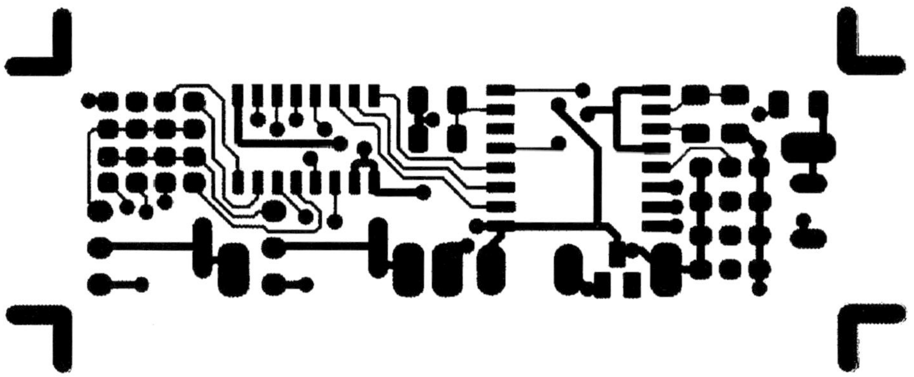

and the holes for mounting traditional components are drilled. After drilling, each hole receives a pipe-formed through-plating, able to conduct electricity. These can be made quite small, a feat difficult to achieve for the hobbyist. If you decide to make the boards yourself, drill the holes afterwards, insert thin pieces of wire in the holes and solder them on both sides of the board. This is the only alternative for attaining the same results as provided by through-plating.

When the circuit board is as small as this one, it is difficult to attain an exact film copy from the paper pages of this book that can be used without reworking. The degree of difficulty increases, of course, when working with double-sided circuit boards, as we have done here. That is when it is good if you can locate films or ready-made circuit boards.

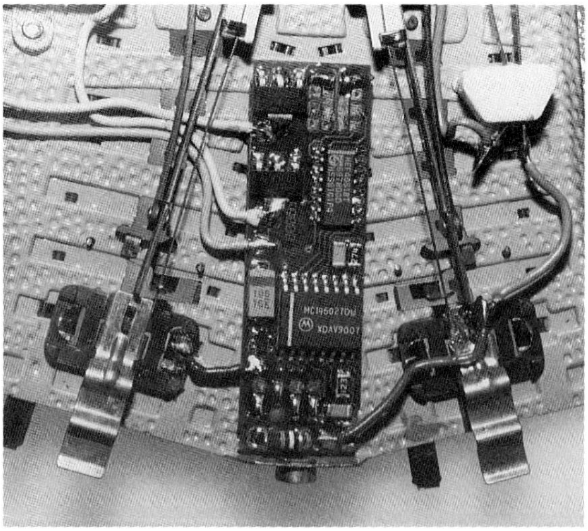

This is the close-up of the wired decoder. Mounting is done via double sided tape.

How to wire the decoder

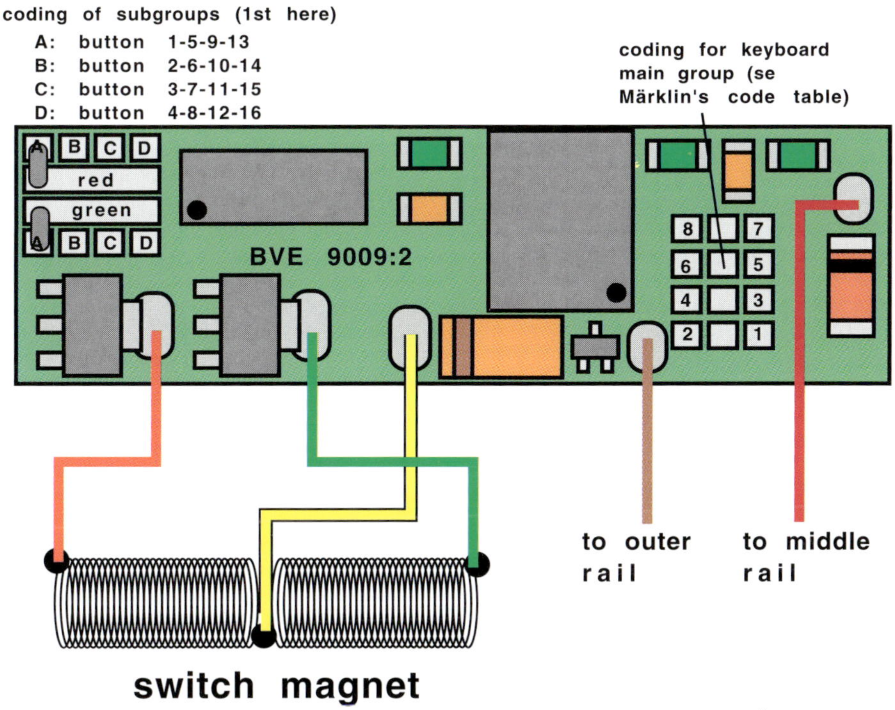

Switch decoder – Schematics

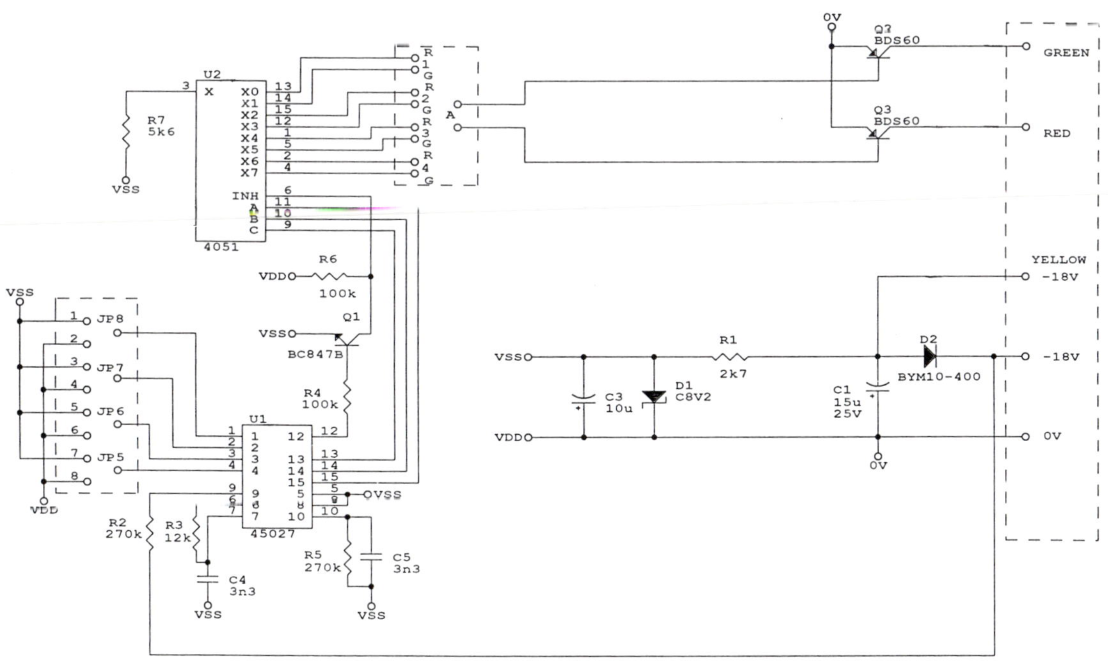

Decoder for Märklin's sheet metal turntable, No. 7186

It is the nature of turntables that you cannot simply connect and control them with a single decoder. The electro-magnetic lock must be activated and the motor must turn in the right direction. Normally, a combination of several decoders is required and control is complex.

This decoder handles all of this. Consequently, the turntable can be operated with a single key action: the green key for clockwise turning and the red key for counter-clockwise turning.

The module is just 4 mm (less than 1/4") in height, making it easy to hide it away under the Märklin turntable's sheet metal base. As seen in the accompanying figure, wiring is simple. If you like, you can perform a function test before installing the decoder using three lamps (Märklin signal lamps). If connected as shown in the test figure, the red and green lamps indicate the turntable's direction of rotation, and the orange lamp, the electro-magnetic lock.

A FEW TIPS ON CONSTRUCTION

Although the circuit board is double-sided, components are only mounted on the upper side (the "component" side). For detailed information on general construction and soldering techniques, see the section on the switch decoder.

Begin assembly by soldering either IC 1's pin 5 or 7 in place. It serves as a centering point. Now solder IC 1's remaining pins and mount all resistors that adjoin the IC. Pin 3 or 4 on IC 2 are both suitable centering points. Finish up assembly in whatever order you prefer.

TROUBLESHOOTING

You will probably be reading this only if the decoder has not functioned correctly during initial testing. Use the test circuit and lamp set-up if you run into problems. It has the advantage of identifying properly functioning circuitry.

Start troubleshooting with an inspection of the circuit board – use a magnifying glass – to detect inadvertent solder bridges and other soldering faults.

Next, check the board's power supply and voltage regulation circuitry. Is a power supply of 18 volts present? Are the ICs supplied with 8.2 volts? Are the voltages too low, too high or just right? The source of power supply faults is usually found at D1, C3 or R12.

With the correct supply voltages, check that the main address and sub-addresses are correctly coded. The solder bridges should have been made as shown in the accompanying illustration. Coding is provided in Märklin's digital instruction book. You'll find the same information in my book on digital model railroading.

If all checks out as described above, you should get indication of some sort from the lamps. Although we haven't inspected the servo section, with its accompanying power transistors, one of the three should light.

Turntable decoder

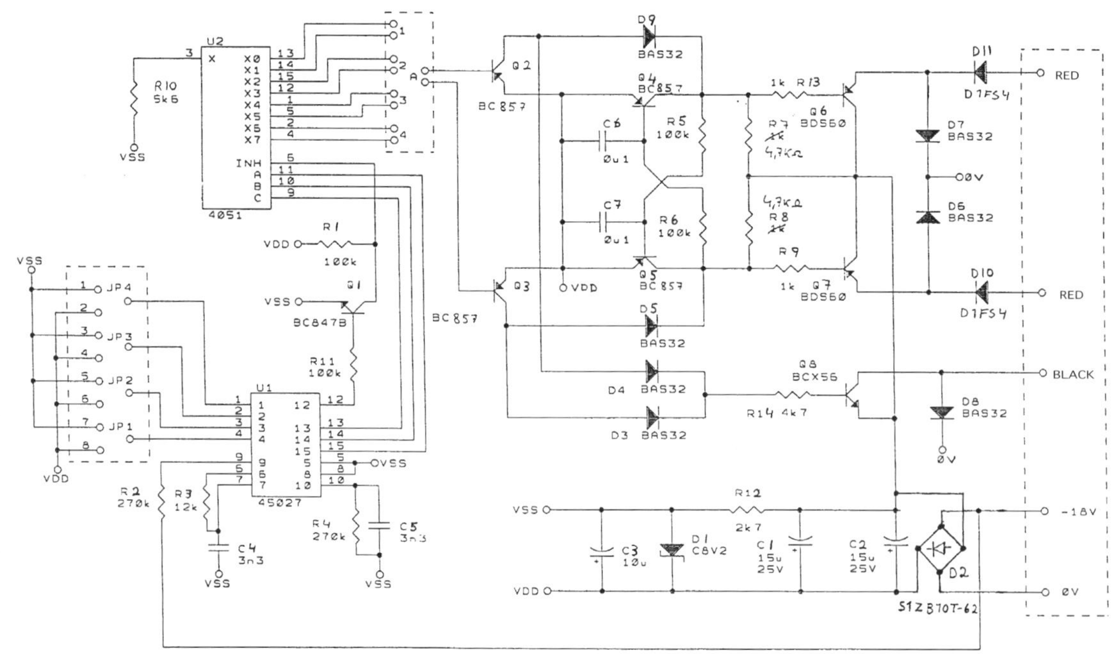

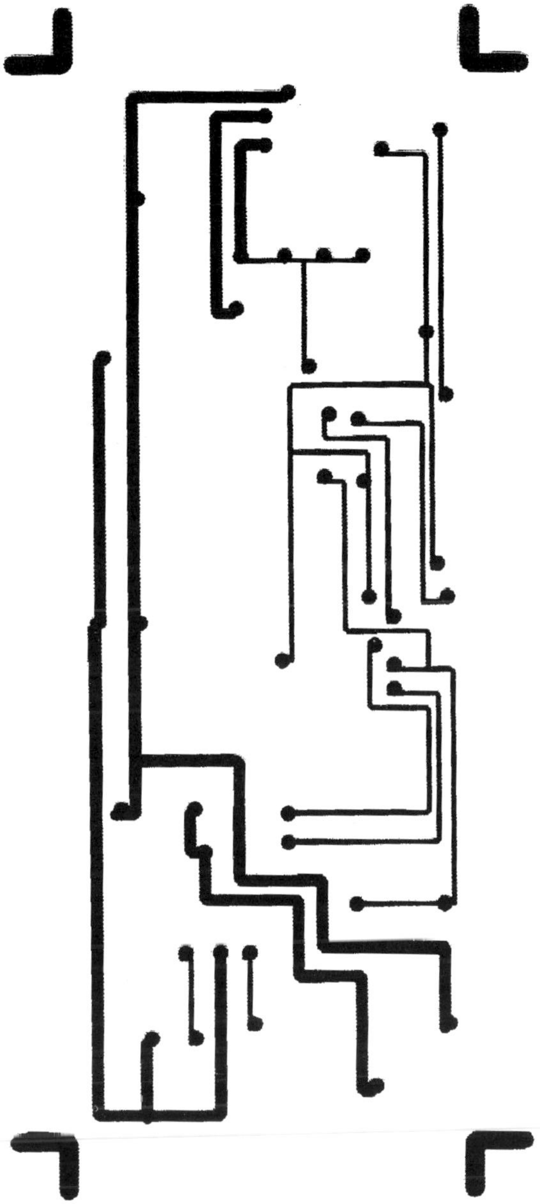

BVE9104
Solder side (scale 2:1)

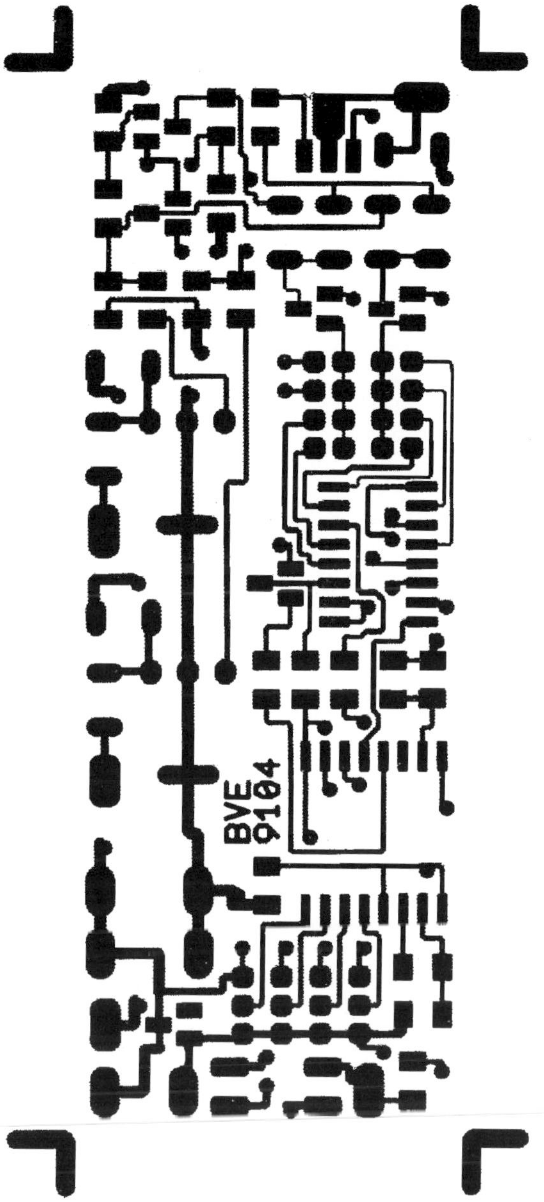

BVE 9104
Comp. side (scale 2:1)

Turntable decoder components list

IC1	45027 SMD
IC2	4051 SMD
D1	Zener diode C8V2
D2	Bridge rectifier S1ZB10T-62
D2	alt. diode BYM10-400
D3-D9	signal diode BAS 32
D10-D11	diode D1FS4
C1-C2	electrolyte SMD 15µF/25V
C3	electrolyte SMD 10µF/16V
C4-C5	capacitor SMD 3.3nF
C6-C7	capacitor SMD 100nF
R1,R11	resistor SMD 100kΩ
R2,R4	resistor SMD 270kΩ
R3	resistor SMD 12kΩ
R5,R6	resistor SMD 100kΩ
R7,R8	resistor SMD 4.7kΩ
R9	resistor SMD 1kΩ
R10	resistor SMD 5.6kΩ
T1	transistor SMD BC 847
T2-T5	transistor SMD BC 857
T6,T7	transistor SMD BDS 60
T8	transistor SMD BCX 56
1x	circuit board BVE 9104

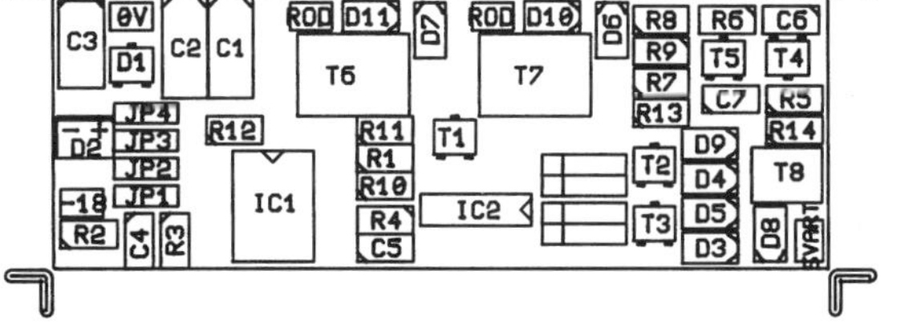

Components layout

Function Test of the Turntable Decoder

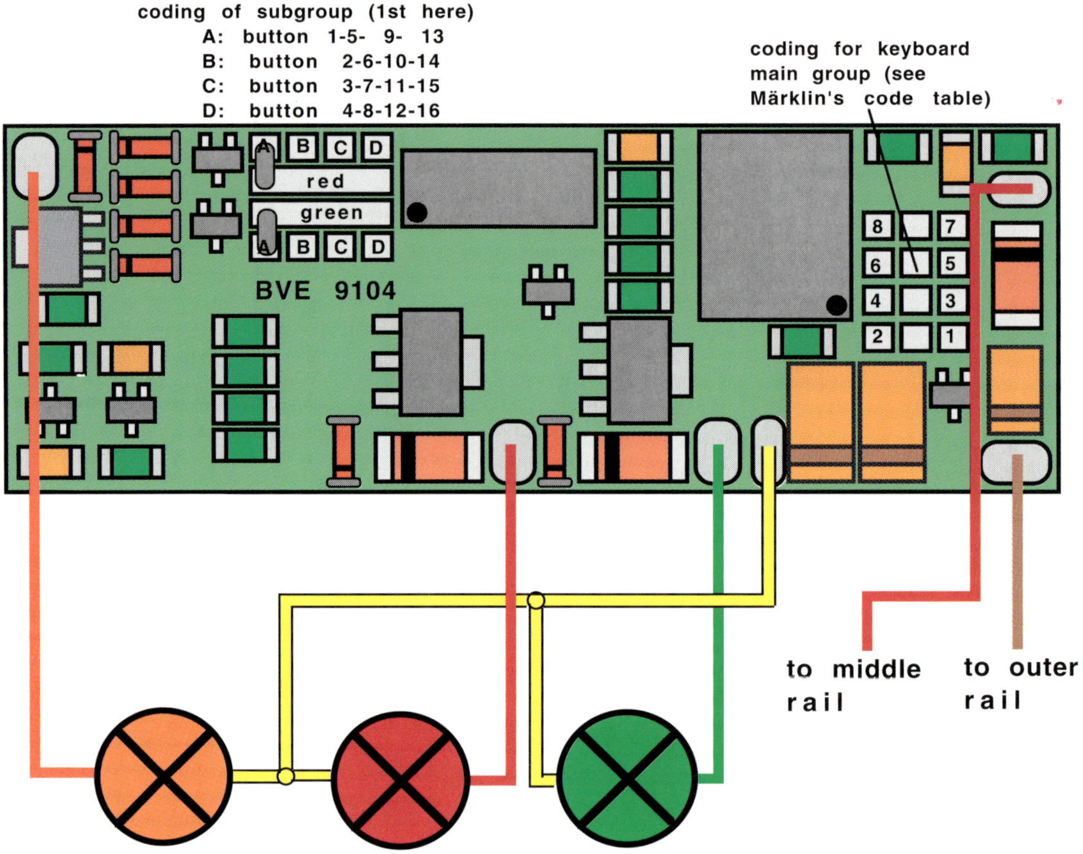

How to wire the turntable

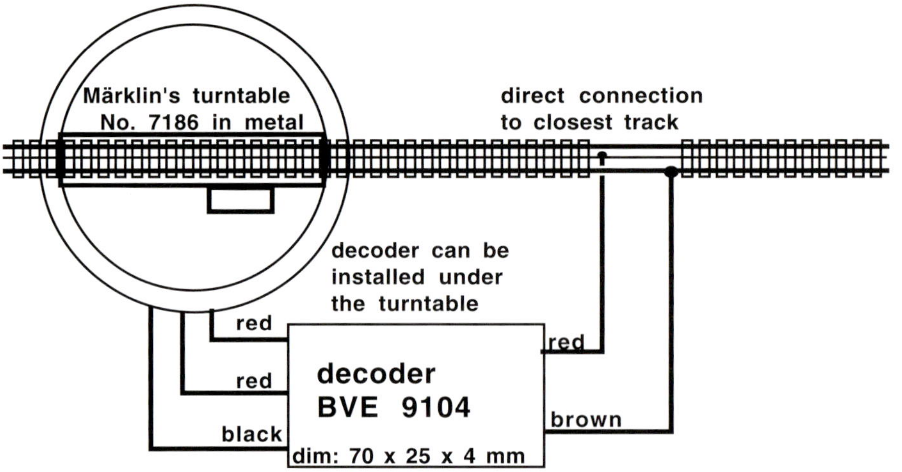

A close-up of the BVE 9104 and mask-painting for easy component assembly. Note how little amount of solder tin is used here.

If you have come this far and still haven't found the fault source, check the reverse side of the keypad. Does it really have the correct group address (1 of 6) set at its quad DIP switch? The first keypad with the lowest address should have all keys in the off position (downwards).

LOGIC PROBES AND OSCILLOSCOPES

For continued troubleshooting, you should have access to a simple logic probe or an oscilloscope. We will first check the signal leads between IC 1 and IC 2. Is a signal present and if so, does it change when keys are pressed? If no change to the signal is noted, IC 1 could be defective. If change is detected, check IC 2's output pins. Two of the four should react when the appropriate key is pressed (red or green). No reaction indicates a possible fault with IC 2.

In principle, if everything is working normally up to this point, the only remaining alternative is that one or more of the transistors are faulty (if the lamps are functioning properly). At least one lamp should light, making it easy to isolate the transistor or transistors not conducting current.

Finally, a tip to keep in mind if a decoder suddenly malfunctions... The usual cause is an overload of some kind. Check the power transistors first.

WHEN WORKING WITH SMALL CIRCUIT BOARDS

With circuit boards as small as this, it is difficult to make an exact and usable film copy from the paper reproduction provided in this book. Naturally, the degree of difficulty increases when using double-sided circuit boards as we have done here. This is when ready-made circuit boards come in handy.

This module is designated BVE 9104. Check the introduction of this book for information on obtaining component kits and circuit boards for this and other BVE projects.

MINITIP

IR transmitter testing instrument

Five active electrical components, a switch and a battery make up this simple test module. It is intended for testing the out-signal from an IR transmitter to a digital control unit but it can also be used, of course, to test other remote control devices (your TV, for example).

The instrument detects invisible infra red light from the transmitter and provides a visible indication. The button is depressed to lead current to both transistors. If IR light strikes the IR transistor's lens (TIL 81 or similar), the transistor will switch on. Via BC 547's base, current is led to the red LED.

The current limiting resistor should be by about 120–150Ω. The 1N4148 diodes is a polarity protector – the circuit is protected if the battery should be connected with reversed polarity.

Mini Tester for the IR-transmitter

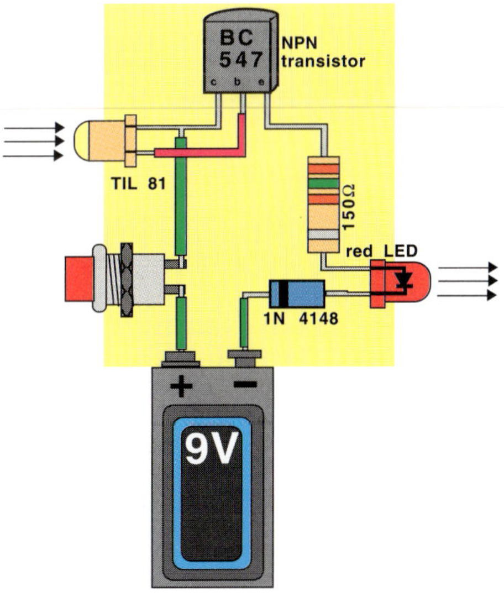

Decoder for Märklin crane, No. 7051

Märklin's remote controlled, sheet metal crane is a classic. It's been around since the late 50's and is on the want-list of nearly all Märklin modelers. The crane stands on a foundation, and the cab and motorized crane arm can turn 360° on a horizontal plane. A second motor raises and lowers the hook. This crane is built for heavy loads. An additional feature is the lifting magnet, controlled electronically by a coil. All electrical components are connected to a special control box via a long cable.

The era of digital electronics has come and that means that even this Märklin crane can be controlled digitally. In passing, there is even a more expensive "big brother" in Märklin's assortment – the digital 7651 crane.

In this section we will build our own decoder, constructed especially for the crane, that controls both motors and the electromagnet. Because our decoder is mounted in the crane's foundation wiring is minimal. Just two leads – a red and a brown – are run from the unit to the nearest stretch of track. Control is via the keypad.

A FEW CONSTRUCTION TIPS

In many ways, construction of this decoder resembles that for the turntable. When it comes to troubleshooting, see the turntable decoder section.

This project is built using surface components. If you have little experience with this type of construction, read the introductory section on tools and soldering. Much of the focus there is on surface mounting technique.

It feels right to start construction on this project by mounting IC 1. I would recommend using pin 4 or 5 as a centering point. Order of assembly after this step is not critical. Just take it easy with the solder so that it doesn't flow to adjoining pins. Even veteran "solderers" sometimes make these unintentional bridges, but they also have the skill to remove excess solder with a "stroke of the pen". This is something that you will learn with experience. One technique is to let the solder overflow cool slightly before "splitting the cells" with the soldering pencil.

SOLDER REMOVAL

There are two ways of removing solder. Both can be good to keep in mind when accidents occur. One way is by using a special suction tool, the smaller the better, to remove excess solder. The other method requires a solder braid, made up of thin, braided copper wire. The braid is positioned over the excess solder and a hot soldering iron tip is placed lightly over the braid. The solder begins to flow and is

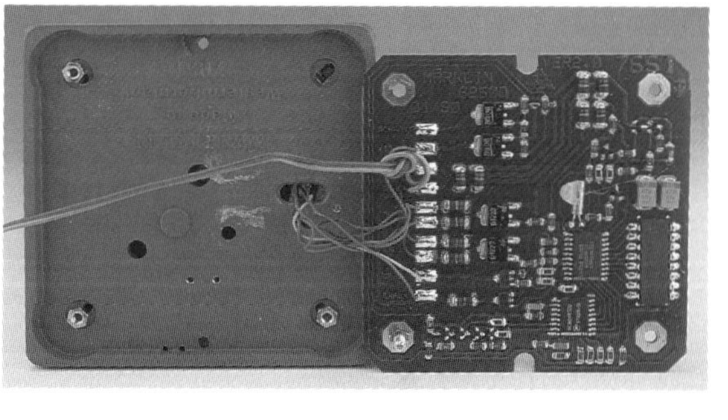

Märklin's crane is an excellent object for digital control. The decoder used was especially designed for crane control. Märklin offers a ready-to-run version but it considerably more expensive. Both variants are mounted in the crane foundation.

Crane decoder BVE9105

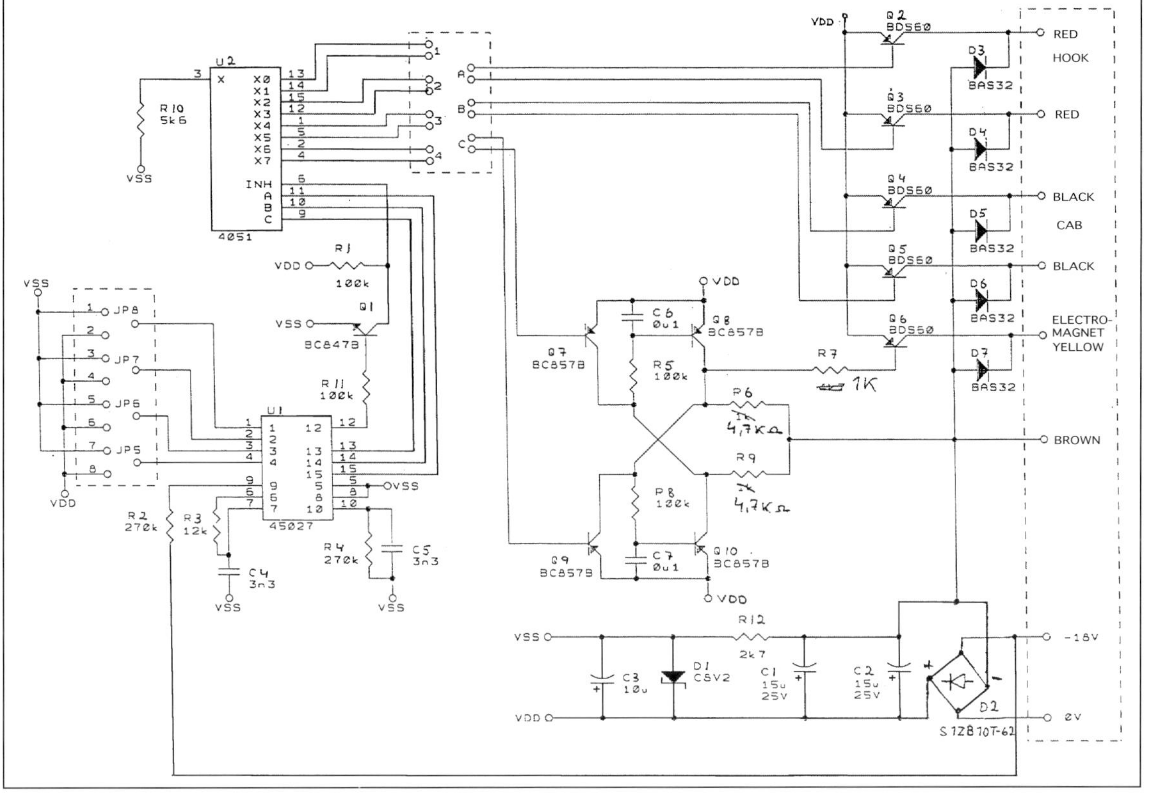

drown up by capillary action into the braid.

The result is a clean surface and ordinary soldering can now be repeated.

These methods are also handy if you want to reprogram a decoder and give in a new address in the digital system.

Previous sections covered circuit board construction in more detail. For this reason, I'll pass over this subject here. But as mentioned before, when circuit boards are as small as the one used for the crane decoder (1" x 2 3/4"), it is difficult to obtain an exact film copy from the book's paper reproductions. Much rework is required. And of course for double sided circuit boards, as we use on this project, the problems are even greater. It is advantageous if ready-made films or circuit boards are available.

This module is designated BVE 9105. Read more in the introduction of this book for sources supplying component kits and circuit boards for this and other BVE projects.

The crane decoder has five power outputs. Two of these control the lifting hook's vertical action and two swing the crane. The fifth controls the electromagnet.

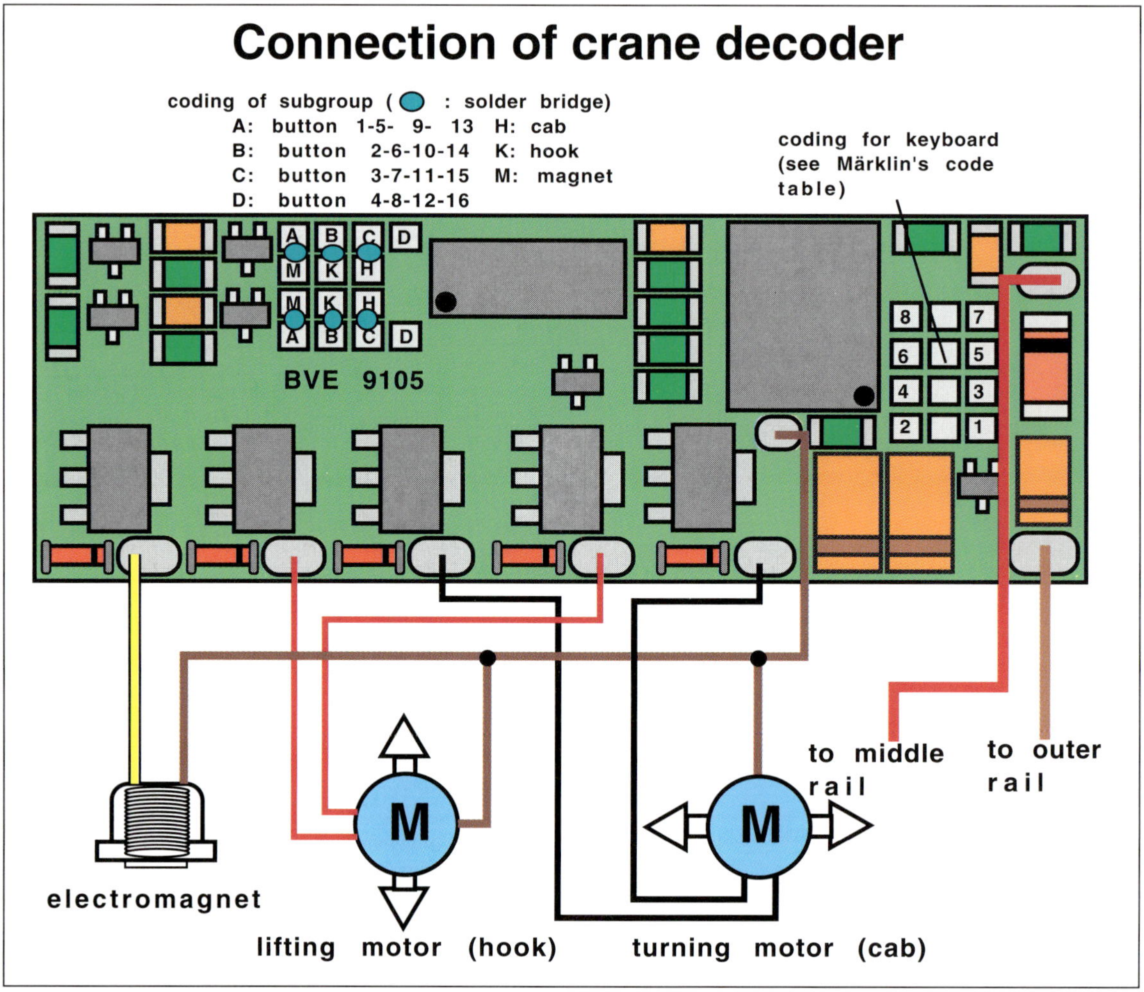

MINITIP

Controlling switches with locomotive detectors

This may seem to be a rather odd tip, but if the day should come when you need a switch decoder quickly, it could come in handy. Using the figure as reference, the circuit lets you can control a switch with a locomotive decoder.

If current (corresponding to full power) is lead in a single direction, one of the two electromagnets will be activated. If the locomotive reverser is thrown and the power voltage increased, the other electromagnet will be activated.

If incorporated into a system, the controller panel can be complemented with a push-button that switches on full power.

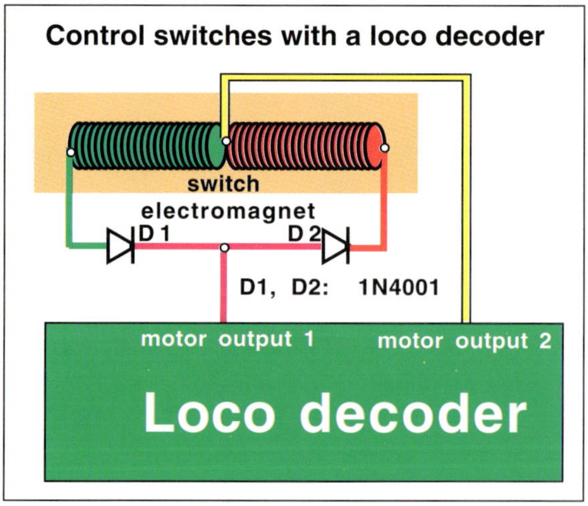

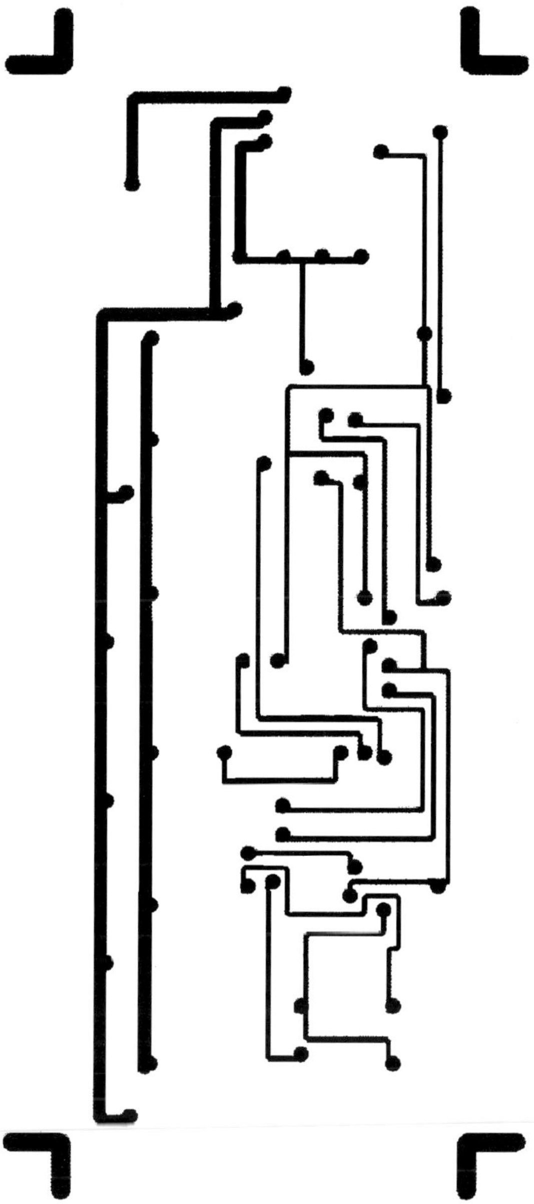

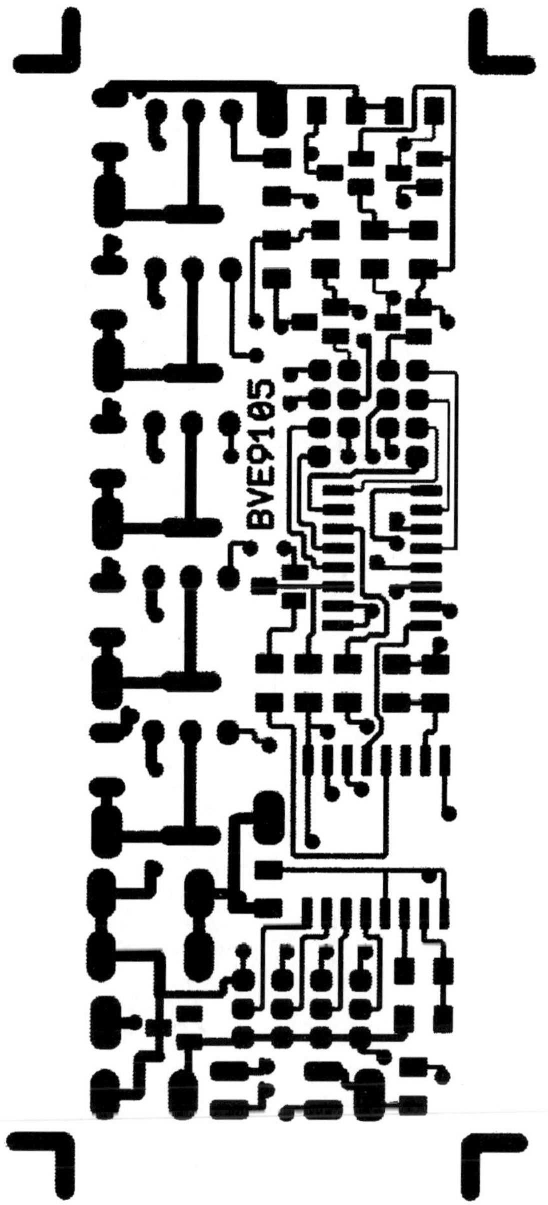

BVE9105
Solder side (scale 2:1)

BVE 9105
Component side (scale 2:1)

Crane decoder - component list

IC1	45027 SMD		
IC2	4051 SMD		
D1	Zener diode C8V2		
D2	bridge rectifier S1ZB10T-62		
D2	alt. diode BYM10-400	R7	resistor SMD 1kΩ
D3-D7	signal diode BAS 32	R10	resistor SMD 5.6kΩ
C1-C2	electrolyte SMD 15µF/25V	R12	resistor SMD 2.7kΩ
C3	electrolyte SMD 10µF/16V	T1	transistor SMD BC 847
C4-C5	capacitor SMD 3.3nF	T2-T6	transistor SMD BDS 60
C6-C7	capacitor SMD 100nF	T7-T10	transistor SMD BC 857
R1,R11	resistor SMD 100kΩ	1x	circuit board BVE 9105
R2,R4	resistor SMD 270kΩ		
R3	resistor SMD 12kΩ		
R5,R8	resistor SMD 100kΩ		
R6,R9	resistor SMD 4.7kΩ		

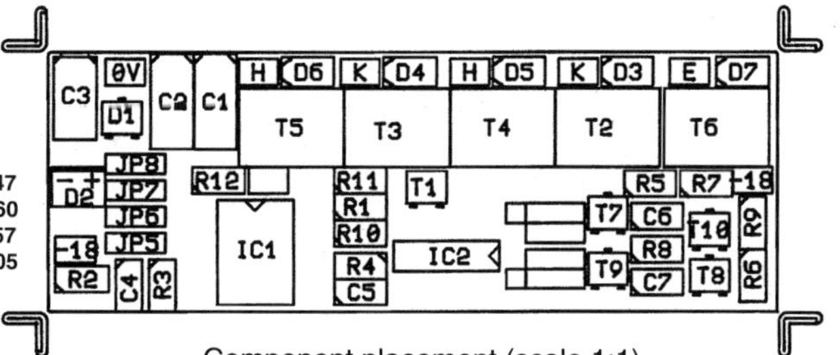

Component placement (scale 1:1)

Decoder for three-way switches

Three-way switches "steal" decoder space and keys from the keypad – at least if you have standard 8-output decoders.

Here is a decoder that is especially designed to regulate three-way switches and has dimensions that are just right (1" x 2 1/2") for installation under the sheet metal roadbed of a Märklin M-switch.

From the operational standpoint, this unit has advantages over standard decoders. To throw the switch points from left to right using standard decoders, you must first set the points to the middle position. With this circuit, you can switch from one extreme to the other with a single maneuver.

If necessary, key positions can be rearranged to match switch positions by altering the module's sub-address solder matrix. In the accompanying figure, I have shown a configuration that could be used for such a subgroup. There are many alternative configurations.

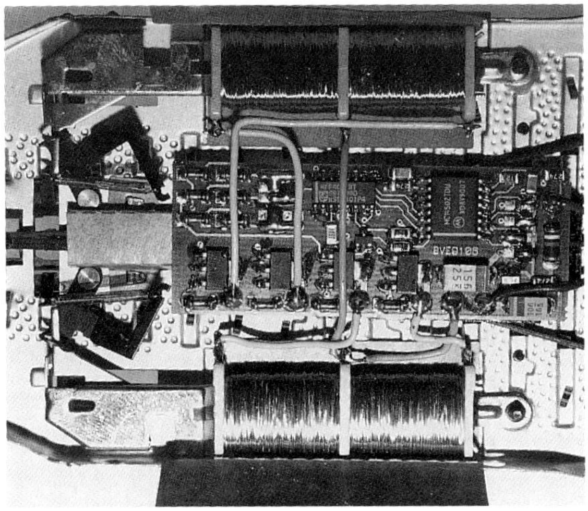

This decoder is especially optimized for remote control of a three-way switch.

A FEW CONSTRUCTION TIPS

This decoder has much in common with the other decoders in this series, especially the circuitry around IC 1 that detects code. Consequently, it is logical that the components have the same values as those used on the other modules. This is an advantage both when buying components and during construction – you are used to working with this assortment.

The power section (the servo) after IC 2 is, however, optimized for each task and differs greatly from other decoders. Even so, you will recognize certain components such as protective diodes and power transistors.

Component mounting procedures are the same as for the crane detector and turntable detector. Here too, you begin construction with either IC 1's pin 4 or 5 as a centering

Switch decoder for 3-way switch

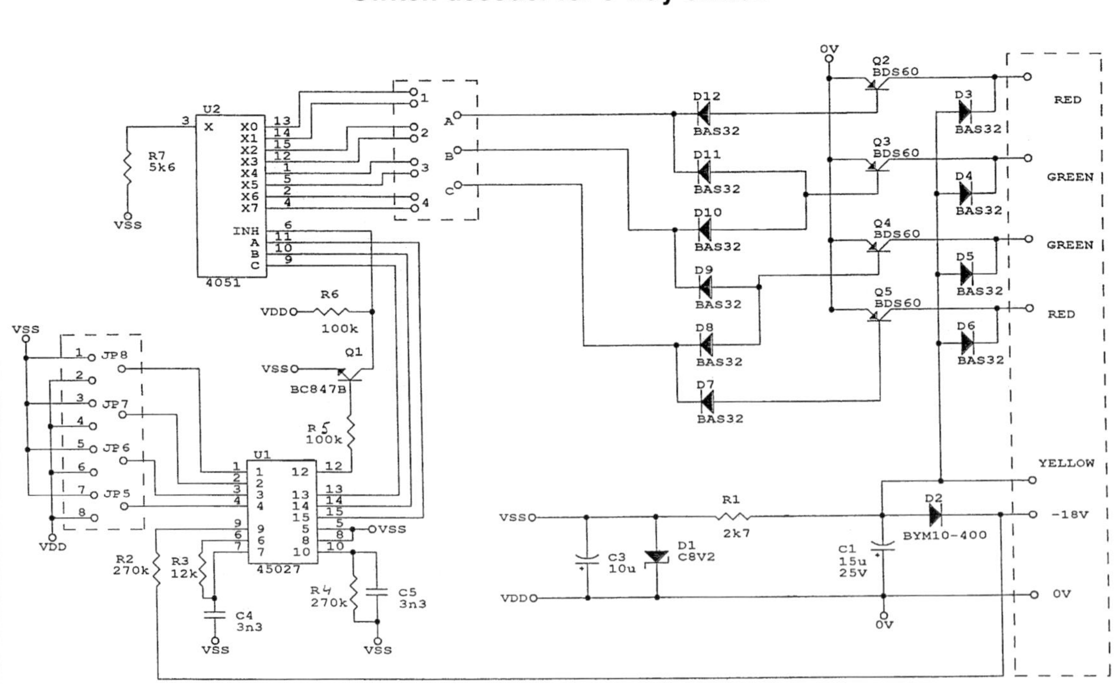

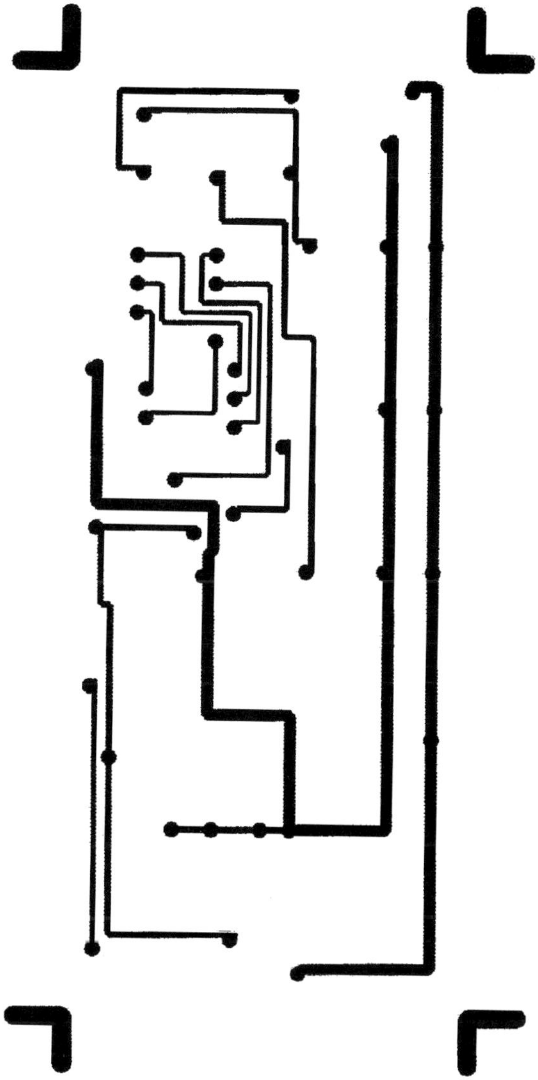

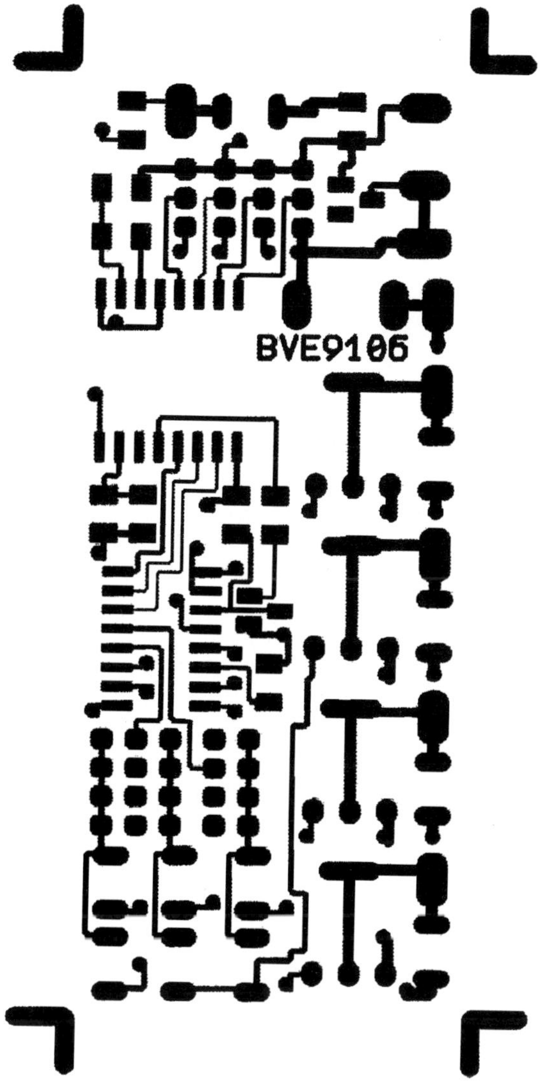

BVE9106
Solder side (scale 2:1)

BVE 9106
Component side (scale 2:1)

3-way decoder component list

IC1	45027 SMD
IC2	4051 SMD
D1	Zener diode C8V2
D2	diode BYM10-400
D3-D12	signal diode BAS 32
C1	electrolyte SMD 15µF/25V
C3	electrolyte SMD 10µF/16V
C4-C5	capacitor SMD 3.3nF
R1	resistor SMD 2.7kΩ
R2,R4	resistor SMD 270kΩ
R3	resistor SMD 12kΩ
R5,R6	resistor SMD 100kΩ
R7	resistor SMD 5.6kΩ
T1	transistor SMD BC 847
T2-T5	transistor SMD BDS 60
1x	circuit board BVE 9106

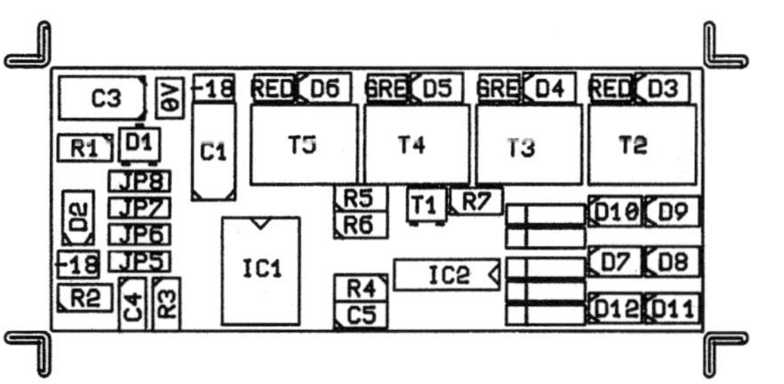

Component placement (scale 1:1)

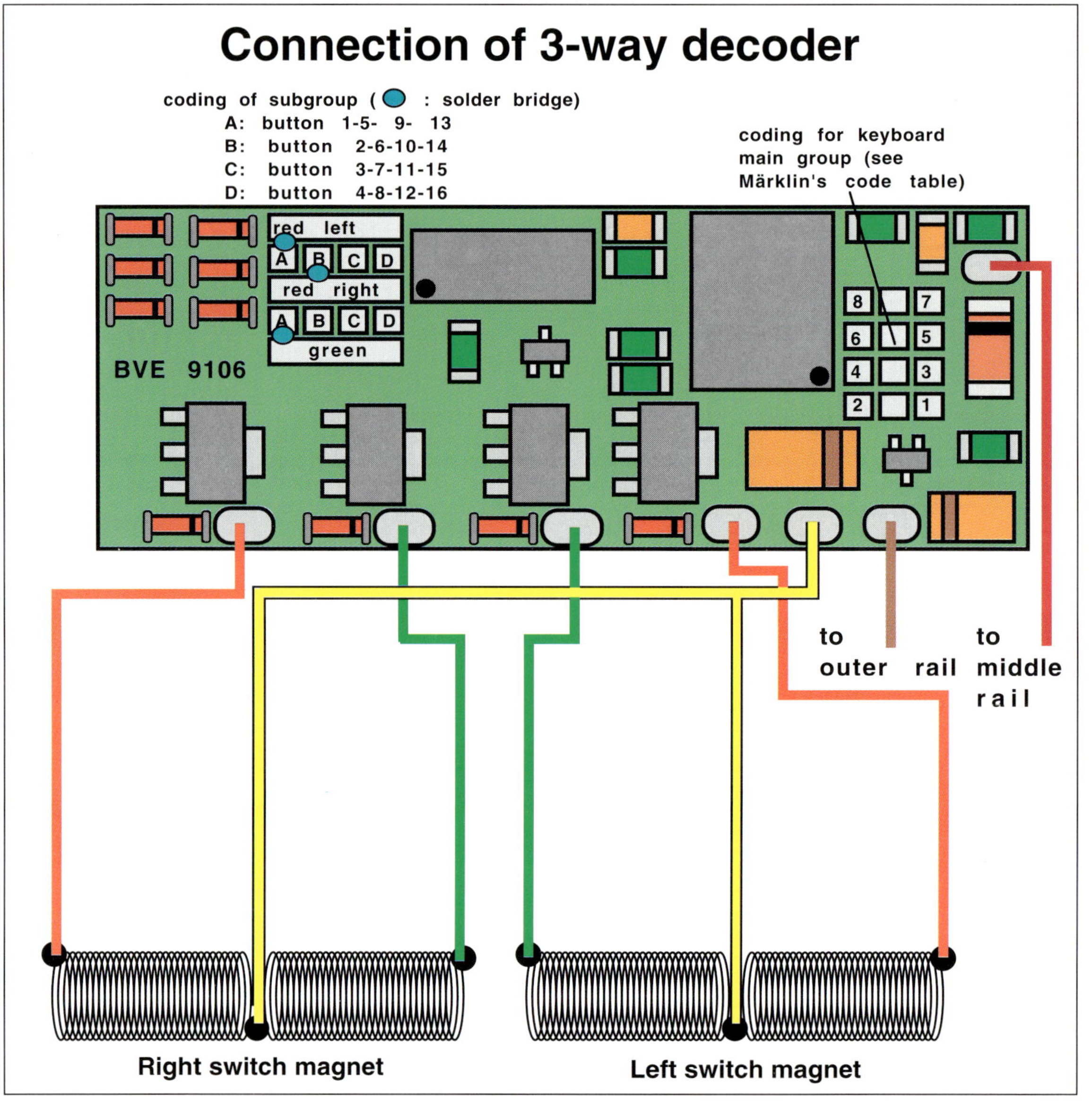

point. If you have already built these other decoders, the remaining construction steps will be easy. We've previously covered circuit board construction, how the decoder works, troubleshooting and repairs.

Now for a few words on decoder installation; let's start with M-switches. Fortunately, all connection points are on the component side. This means that we can mount the decoder before making the electrical connections.

For mounting the decoder I like to use thick, heavy duty double-sided tape it it's available or, alternatively, a glue gun. With few drops of melted glue, you can position the decoder a tad from the mounting surface (to avoid shorts) and you'll also be able to remove the detector if repairs must be made on the board's solder side.

With K-switches, you don't have the option of hiding the decoder on the switch itself. On the other hand, K-track is normally used differently than M-track. K-track is often used with more advanced layout construction. The decoder can just as well be mounted under the benchwork. And this is where glue guns are especially useful.

As far as wiring goes, I usually hook up the power feed first and test the decoder without a load. Careless errors are detected without damage to components caused by excessively high currents. Using this method, and with a small logic tool, much of the circuitry (including output functions) can be checked. Finally, the load is connected.

The module is designated BVE 9106. See the book's introduction for information on obtaining component kits and circuit boards for this and other BVE projects.

MINITIP

Coach lighting for AC operations

There are countless solutions for the perpetual problem of lighting passenger cars. Here is a solution that even takes care of end of train lighting. The interior ceiling lamps are actually high-intensity LEDs. You can also model the table lamps in a diner. Two of the lighting sources on the schematic represent table lamps. The transistor circuitry functions as a current limiter. It provides the advantages of starting at very low voltages and supplying a constant illumination level, regardless of train speed (track voltage).

The recommended total for parallel wired LED-resistor-LED units is not written in stone but should be between two and five.

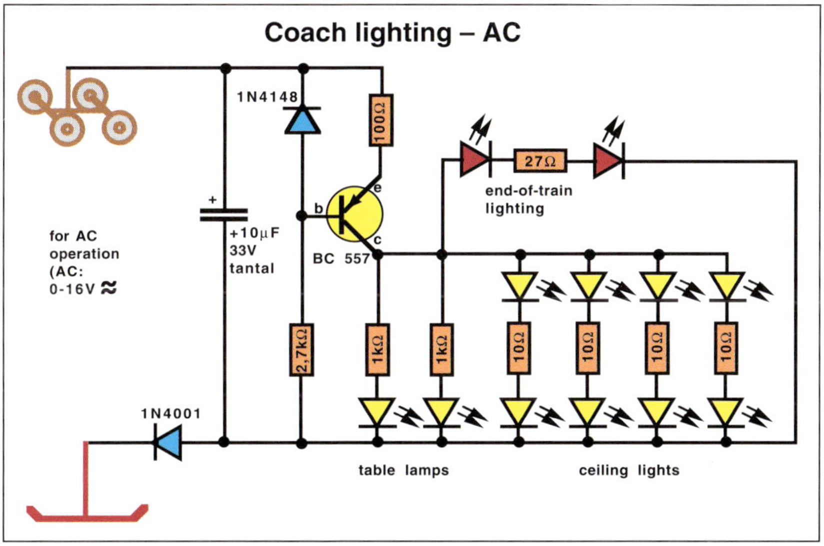

MINITIP

Coach lighting for N and HO scales

This is a universal lighting system that works with both AC and DC. The circuit uses LEDs as lighting sources. High or low intensity LEDs are used depending upon how they are placed – as table lamps or ceiling fixtures.

The figure's outer rectangle (with green and red leads, and four diodes) is a bridge rectifier with the AC inputs connected to the coach's current collector.

Regardless of the drive system or direction of travel, lighting will switch on as soon as four volts are present at the rails. An electrolytic capacitor of 220 μF supplies power during momentary interruptions common when passing switches and block divisions as long as train speed is not excessively slow.

Current reducing resistors of 220 Ω can be exchanged for 330 Ω resistors if digital operations are used with voltages above 16 volts.

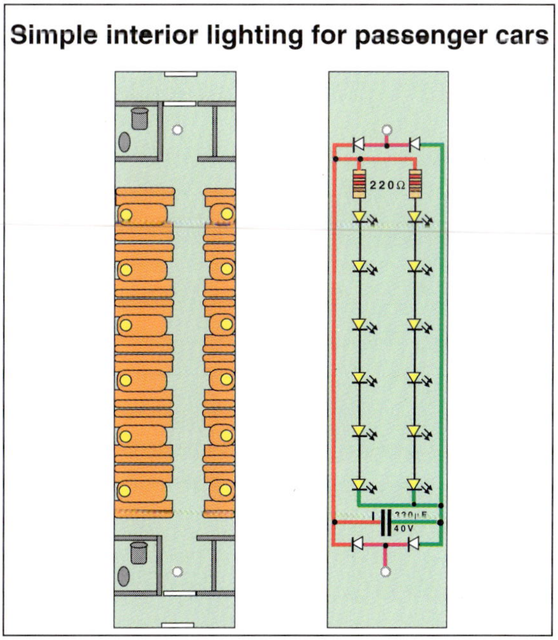

F1–F4 function decoder for engines and cars

Märklin's C80 and C90 series locomotive decoders have an accessory function, F0, for remote headlight control. There are also buttons on the controller for an additional four functions, namely F1 to F4.

Now days, there are considerably more features on engines than just motors and headlights that can be controlled remotely. Examples are cab lights, smoke units, Telex couplers and pantographs.

Function decoders for keys F1 to F4 can also be useful for coaches (interior lighting and end of train lighting) and special rolling stock such as cranes and cement mixers. They could even be used to open and close box car doors.

This function decoder corresponds well to the control requirements of these applications and is also quite compact – 18 x 7 x 6.6 mm. You can draw 200 mA from each output and if you like, use them all at the same time.

It is often best to code the function decoder with the same address as the engine's ordinary decoder. This makes all functions available at one address in that F1–F4 are supplementary functions.

If you need even more functions, a second module can be connected with a different address.

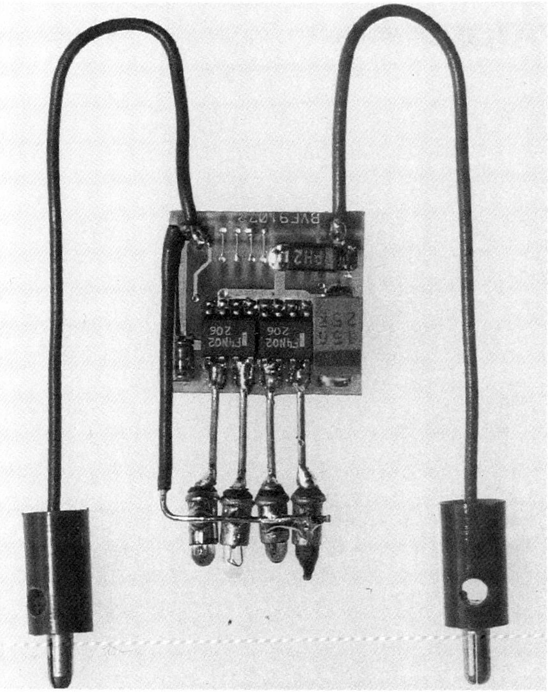

The usual procedure is to use the same address for the function decoder as for the locomotive.

A FEW PRACTICAL CONSTRUCTION TIPS

Despite its smallness and the relatively few components used, this is not a project for beginners. Its compactness makes it rather advanced.

Both sides of the circuit board are used for components.

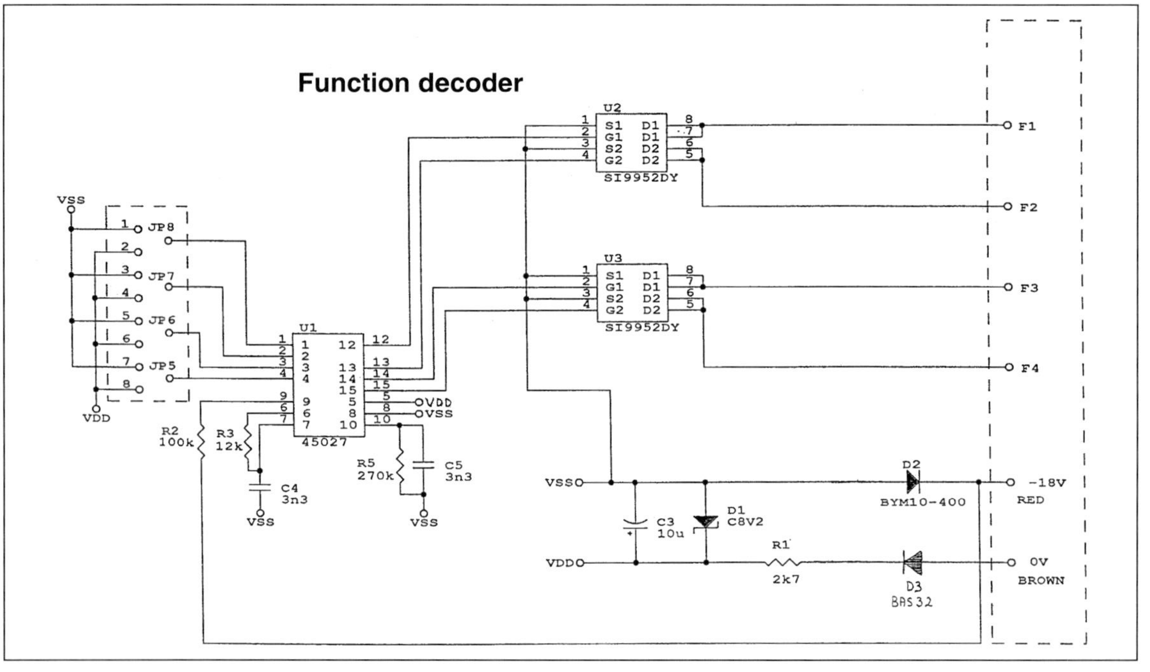

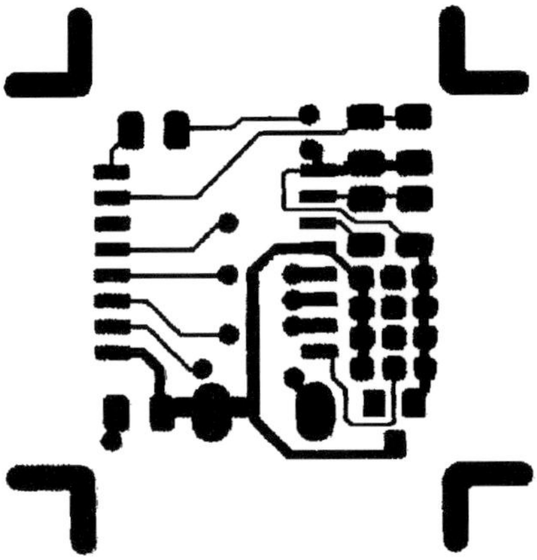

Function decoder component list

IC1	45027 SMD
IC2-3	MMDF4N02 SMD
D1	Zener diode C8V2
D2	diode BYM10-400
D3	signal diode BAS 32
C1	(electrolyte SMD 15µF/ 25V)
C3	electrolyte SMD 10µF/16V
C4-C5	capacitor SMD 3.3nF
R1	resistor SMD 2.7kΩ
R2	resistor SMD 100kΩ
R3	resistor SMD 12kΩ
R5	resistor SMD 270kΩ
1x	circuit board BVE 9107:2

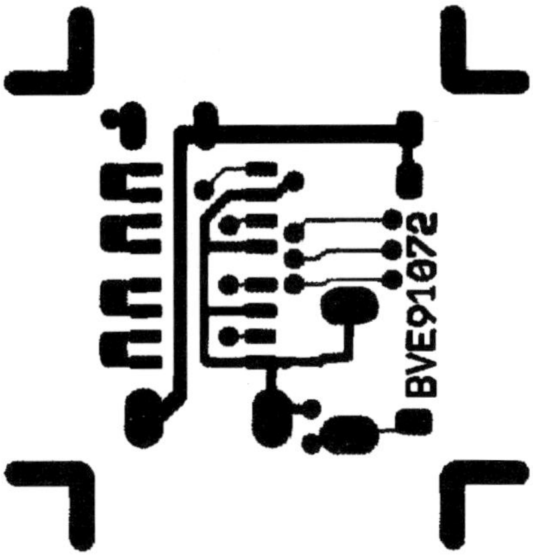

BVE 9107:2 Component side (CPU), scale 2:1

BVE 9107:2 solder side, scale 2:1

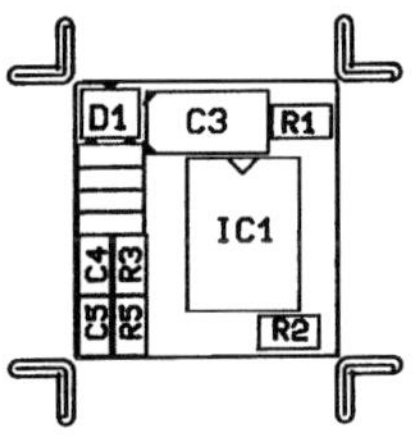

Component placement on component side (scale 1:1)

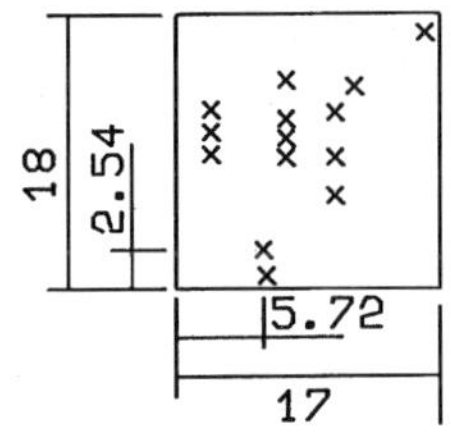

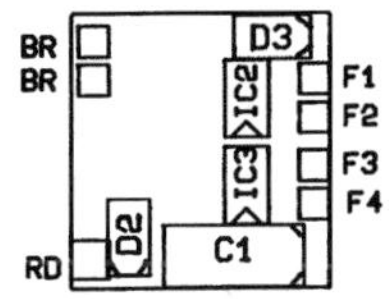

Component placement on solder side (scale 1:1)

Troubleshooting the Function Decoder

Under certain unfavorable circumstances – although very rare – one or more of the BVE 9107 function outputs may switch on by itself.

For this reason, it can be good to be aware of the following. This fault can arise only when the following circumstances occur simultaneously.

1- BVE 9107 is connected to the same collection shoe as the loco decoder.
2- The address is the same as that for the loco decoder.
3- The locomotive's F0 function is activated.
4- The loco is running at maximum speed.
5- All functions F1-F4 are off.

If all of these are present, eliminate the risk for problems by installing a 33 KΩ/1/8 W resistor between pins 8 and 9 on IC1 45027.

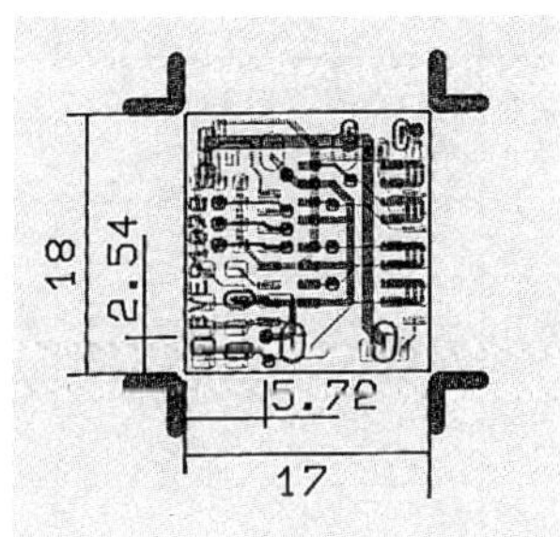

It is best to start construction with IC 1, using one of its pins 5–7 as a centering point. Solder the remaining IC pins before moving on to the resistors and Zener diodes on the same side of the board. This side is completed with installation of the electrolytic capacitor and coding of the address on the code matrix. Although the latter is extremely small, the same principles as for the other BVE decoders apply. Count from left to right on the matrix.

TRANSISTORS IN IC CASES

On the solder side, there are also spaces for a number of components besides the wiring connections. First, IC 2 and 3 must be mounted. These are actually power transistors – N-channel MOSFET in SO-8 cases. Here it can be appropriate to start with pin 2 or 3 and as a counter weight to centering, one of the double pins 5-6 or 7-8. After this, you can solder the remaining IC pins. Now the hard part is behind us and the last three components are easy to solder in place. Connections are shown in the accompanying figure.

Sometimes you may wonder which pin is which on a SMD IC. The rules, or rather praxis, are somewhat different from those for the "heavy weight class" of hole components.

On some ICs you will find a white, round point closes to the first pin. And sometimes even a white dash at the IC end to your left when pin 1 is at the lower left corner (closest to you). Others have a small notch in the case at the left end.

Don't assume that designation texts are right side up. I've been fooled a number of times by upside down texts

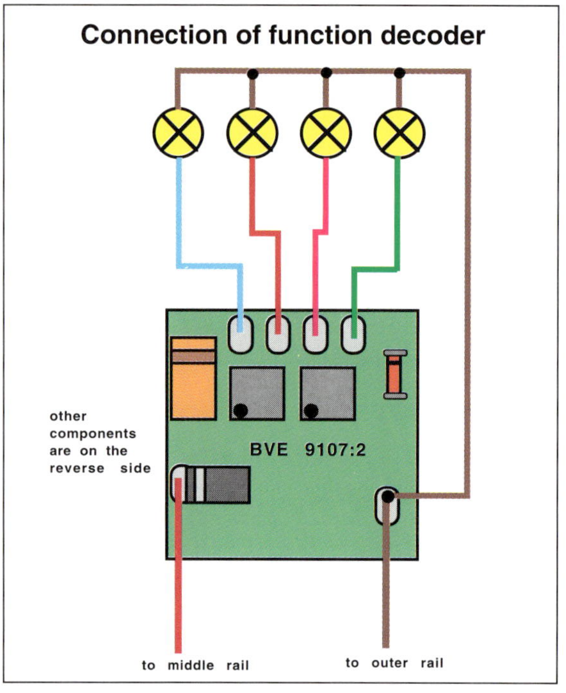

and the result has been destroyed ICs.

DECEPTIVE ELECTROLYTIC CAPACITORS

A diode usually has a bar marking of some sort at its cathode corresponding to the diode symbol bar. Electrolytes, on the other hand, can be tricky. Most often, you will have to check the manufacturers' data sheets to find out which pole should be connected to plus. Contrary to hole components, the bar is normally at the plus end on surface mounted components. Certain manufactures have are considerate enough to add a plus symbol in relief at the bar.

Finally, a few words on films and circuit boards. Information on ready-made films and circuit boards can be found in the book's introduction. This function decoder is designated BVE 9107:2.

MINITIP

DC controller without PWM pulses

Nearly all controllers "of class" have PWM (Pulse Width Modulation). Shorter or longer, full-voltage pulses are sent out to the engine for starts and slow running, or for running at speed – the shorter the pulses, the slower the speed.

There are increasing numbers of motors used for model railroading that "don't like" PWM. These are the very small motors and those with air-cored rotors. Faulhaber and Escap are two suppliers of such motors.

This circuit filters controller output voltages. It can regulate both speed and direction of travel. A high wattage automotive lamp provides simple overload protection. Normally, current passes through the lamp to the locomotive without so much as a soft glow. When an overload is present, the lamp lights.

Note that the LM 350 voltage stabilizer must be equipped with a heat sink. This has to be insulated from the system ground to avoid a short at the speed potentiometer. The in-line diodes allow you to drop the output voltage to zero volts so that even the most sensitive motors come to a standstill.

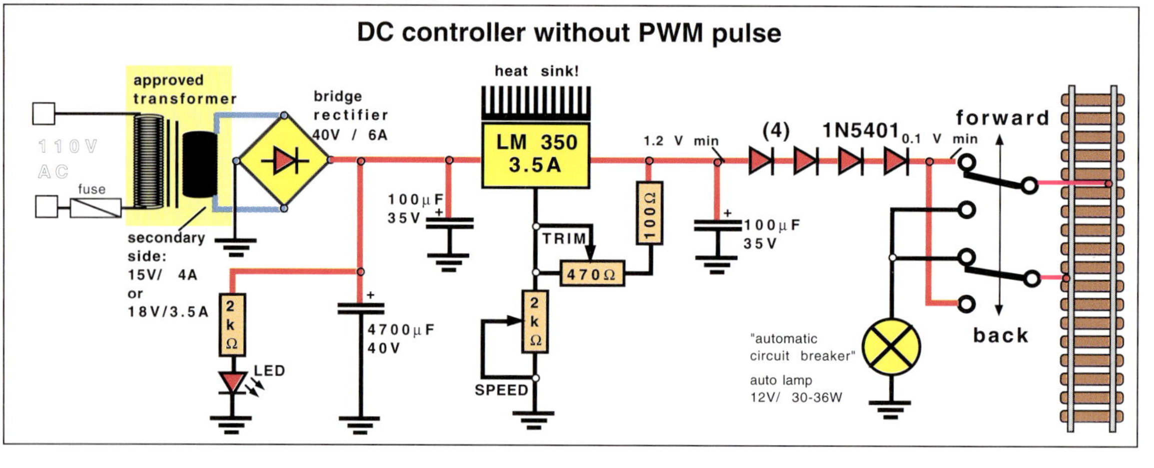

Detector with bistable outputs

As you can indirectly deduce from the heading, most decoders for remote accessory control have monostable outputs.

Monostable function in this context means that the output is active for as long as a command to the output is transmitted from the command center. When the instruction (command) breaks, the output is deactivated. A monostable circuit may include a supplementary timer function that keeps the output active for a specified period of time after the instruction break.

The term bistable means that there are two stable states. A short command is given for activation and the output retains this state even after the command ends. With a dual-position electromagnetic relay, the output keeps its function state both after the instruction terminates and when the layout is shut down and restarted. Only when a new

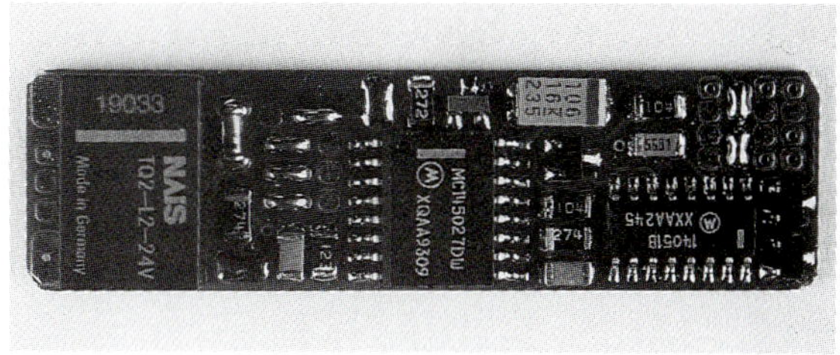

command is given does the output switch to its other stable state.

WHY IS A BISTABLE OUTPUT GOOD FOR DIGITAL OPERATIONS?

After so much talk of bistable functions, there must be

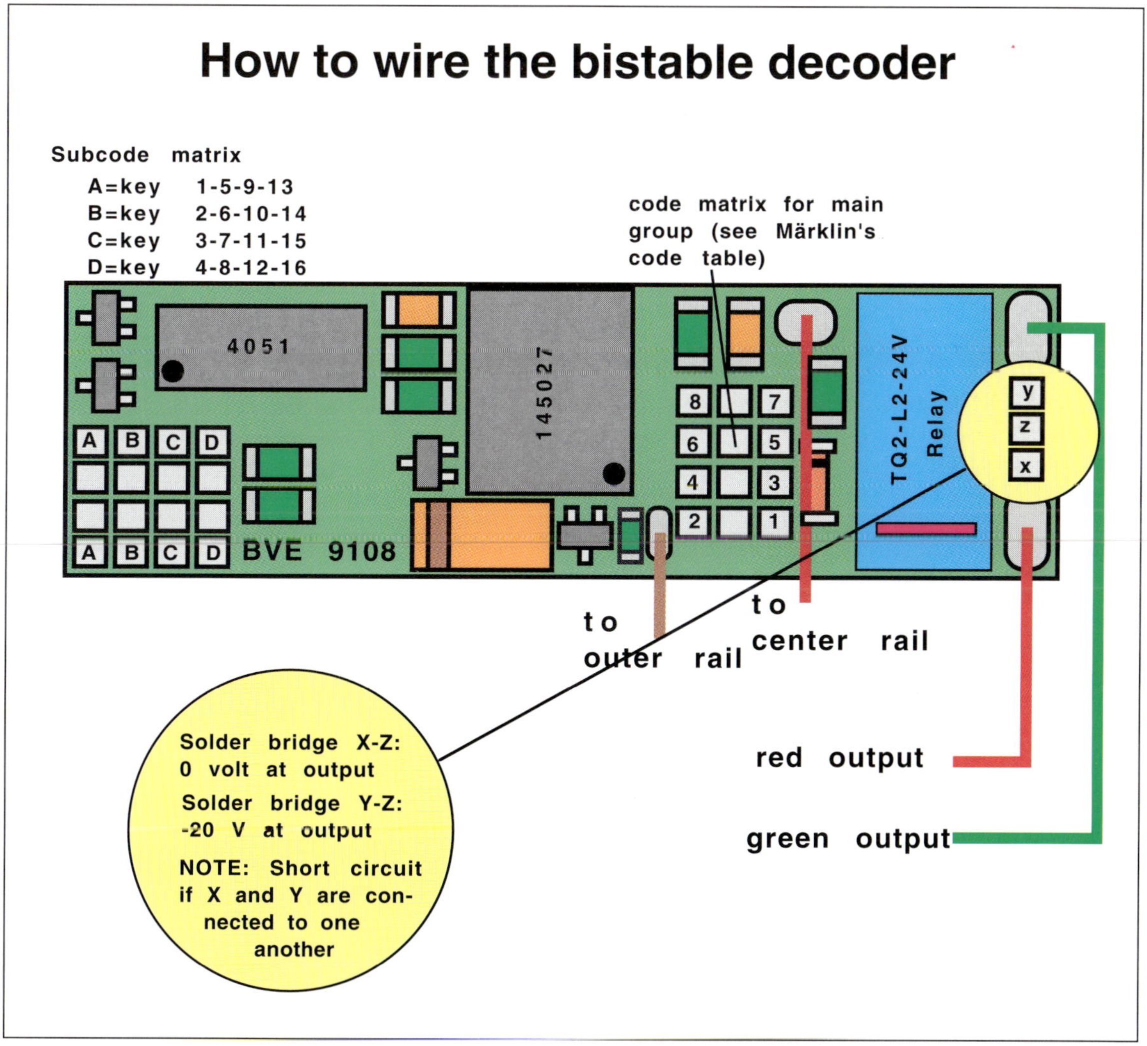

Bistable decoder

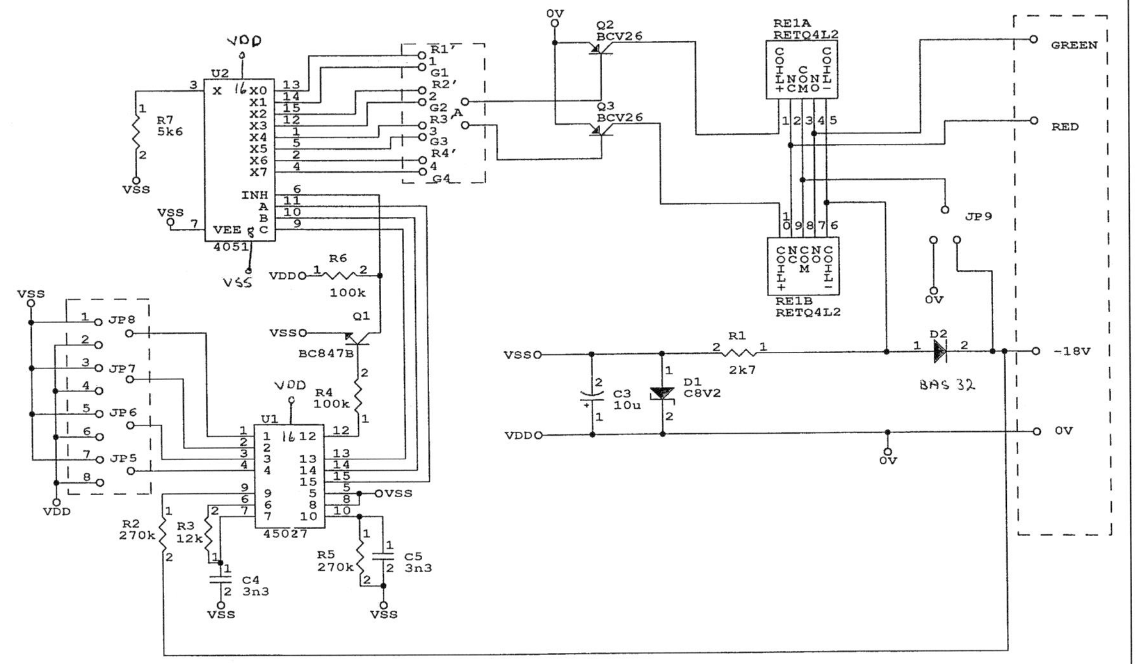

some finesse with it. Let's take a quick look at how instructions are distributed on the digitally operated layout. Via a central unit, virtually the entire layout is wired through a single, 2-conductor output cable to the track. This is the distribution path for electrical power and all instructions, in serial form, to the engines, switches, etc.

We'll use Märklin's crane as an example. We're lifting scrap iron with the electromagnet. With our traditional monostable function, the load is left hanging as long as the right key is depressed on the keypad. That's fine up to this point. But a consequence of this is that all other commands to the layout are blocked. We can't even start the motor that raises the load, not to mention locomotives, etc. Our only alternative is to release the key and drop the load.

This is where a bistable function comes in handy. We activate the output relay at the decoder so that continuous current is delivered to the crane's electromagnet. We can now give other instructions without releasing the load. When the crane has been maneuvered to the proper position, we can drop the load with a second instruction to the relay.

This is just one of the many situations that make bistable functions invaluable.

A FEW CONSTRUCTION TIPS

On this circuit board are several current paths and coding options that should be noted. There is the usual group address coding and sub-address coding at the IC 2 (U2) output pins. As described in detail in a previous section, there are four different sub-address options for the decoder. This means that you can connect four identical decoders with the same group-address and still have different sub-addresses accessible by way of specific red and green keypad keys.

Another option is shown on the schematic. It is the JP 9 bridge, where you determine if the relay contact shall be connected to "plus" or "minus", as related to power feed. Note that here we assign the highest potential to zero volts and the lowest to minus eighteen volts.

Naturally, if we should require another adjoining bistable relay, there is always the option of constructing an additional decoder. But with a little improvisation, you can avoid unnecessary expenses by connecting to the original decoder. On IC 2 are three unused dual outputs, for a corresponding total of bistable relays.

Two additional BCV 26 transistors are needed besides the extra relay. Although I haven't done this, it would be good to solder the transistor collector leads (c) to the relay coil's plus pole to physically mount the respective transistors. To shorten leads, a further refinement would be to mount the extra relay as close as possible to the board's ordinary relay – perhaps even gluing the relays together, back to back.

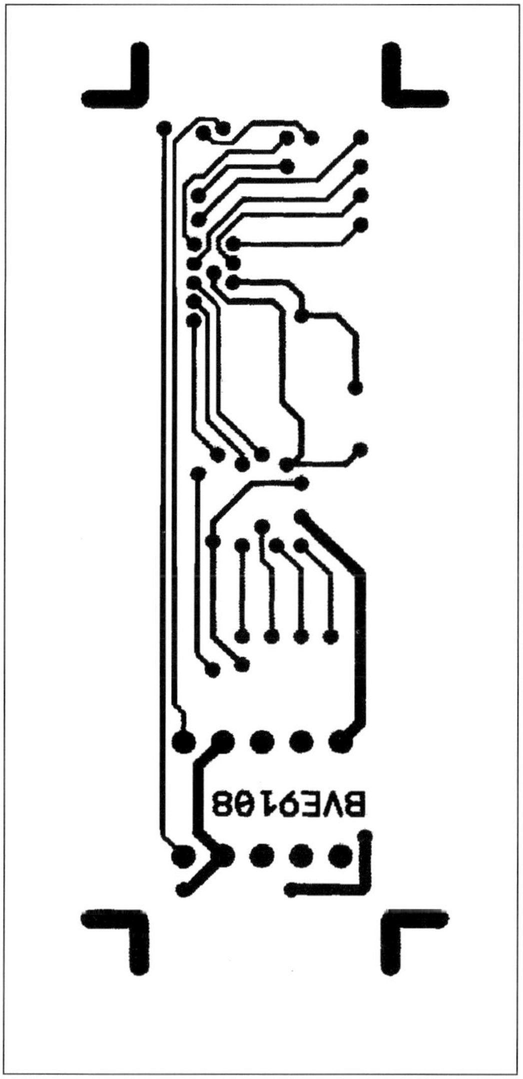

BVE9108
Solder side (scale 2:1)

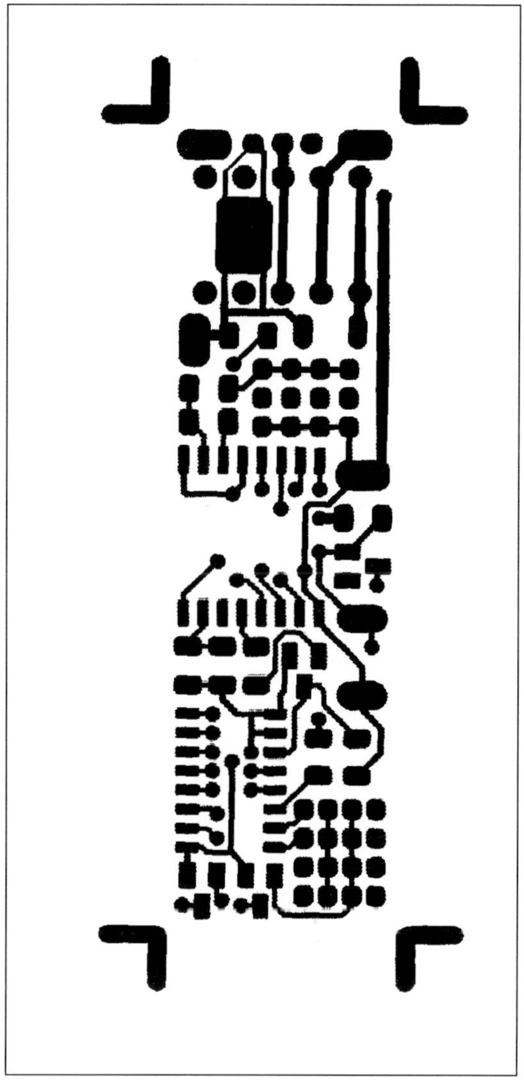

BVE 9108
Components side (scale 2:1)

Component list
Decoder with bistable output

IC1	45027 SMD
IC2	4051 SMD
D1	C8V2 Zener diode SMD
D2	BAS 32 power diode SMD
T1	BC 847 transistor SMD
T2-T3	BCV 26 transistor SMD
C1-C2	not used
C3	10µF / 16V electrolyte SMD
C4-C5	3.3 nF capacitor SMD
R1	2.7kΩ resistor SMD
R2,R5	270kΩ resistor SMD
R3	12kΩ resistor SMD
R4,R6	100kΩ resistor SMD
R7	5.6kΩ resistor SMD
RE1	RETQ4L2 bistable relay SMD
1 x	circuit board BVE 9108

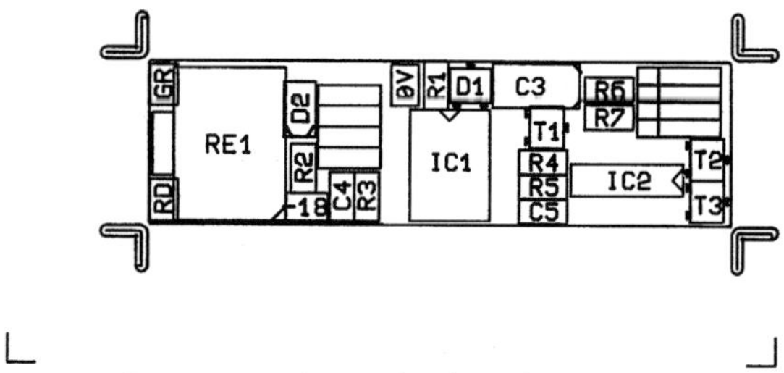

Components layout (scale 1:1)

Flashing end-of-train lighting for Märklin systems

This module is primarily intended for controlling the flashing warning lights on a train's last car. The lamps (LEDs) blink simultaneously at approximately the same rate as used by the Swedish State railways.

Two LEDs (normally 3 mm) can be connected to the module, but a single LED may also be used. Power feed is from the rails via the car's wheels or collection shoe.

The circuit is designed for both Märklin's 3R digital and standard analogue systems (AC).

There is a variant of the circuit detailed in one of the accompanying figures for using the module to control grade crossing warning lights. Here, the LEDs flash alternately

The two electrolytic capacitors are the largest components on this flash module. IC 4001 should be mounted first. This project is relatively easy to build, with few components.

A FEW CONSTRUCTION TIPS

Because the module is built with surface components, you may be inexperienced with the essential construction methods. This is a relatively good project for beginners – components are few and with plenty of space around them. If you haven't worked with SMDs (Surface Mount Devices) before, see the introductory chapter on tools and soldering first.

Begin construction with pin 10 or 11 on IC 2. From a soldering standpoint, these are a little harder to get at than the others. Solder one pin first and use it to center the IC. If you notice that the IC is slightly out of position, it is still rather easy to re-heat the solder with the soldering pencil and adjust the IC's position. As soon as the second pin is soldered, adjustment becomes much more difficult.

Continue by soldering all pins on ICs 1 and 2. After this, you can solder the remaining low-profile components, i.e., all components except for the two larger electrolytes, which are soldered last.

The module is designated BVE 9101. Read more in the introduction on how you can obtain component kits and circuit boards for this and other BVE projects.

Flashing end-of-train lightning Components list

IC 1	7805, SO-8
IC 2	4001, SO-14
D1	(4) BAS 32 (equivilant to one bridge rectifier)
R1	220 kΩ SMD
R2	2.7 MΩ SMD
C0	6.8µF / 35 V SMD
C1	68µF / 6V SMD
C2	220NF SMD
(1)	Circuit board BVE 9101

Flashing end-of-train lighting - schematics

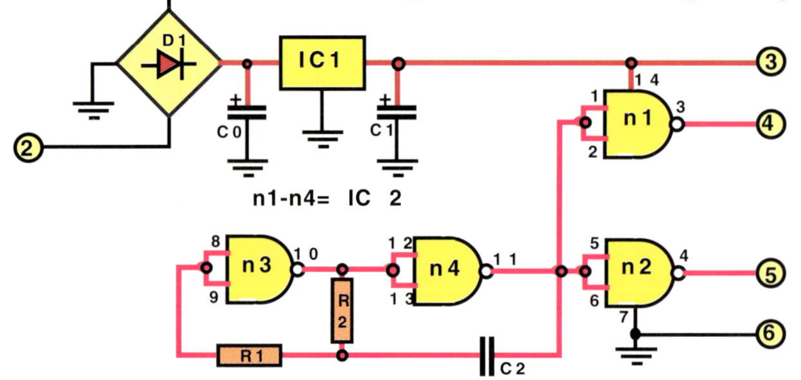

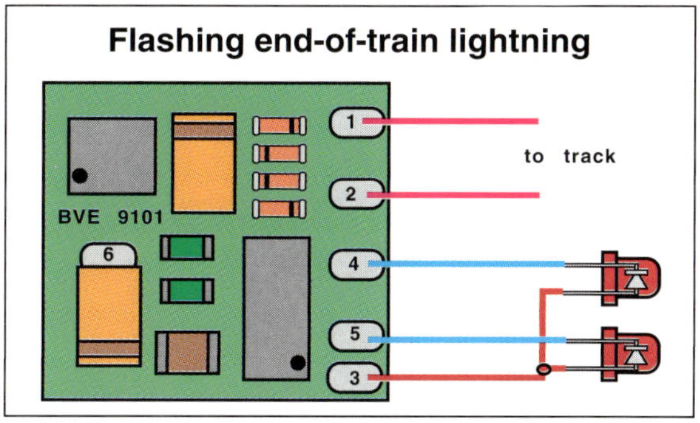

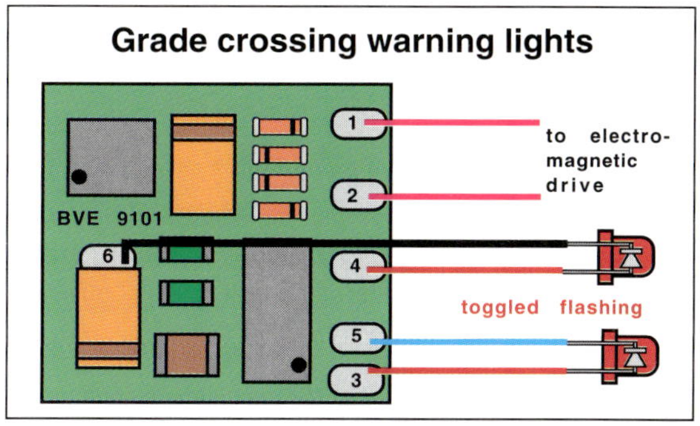

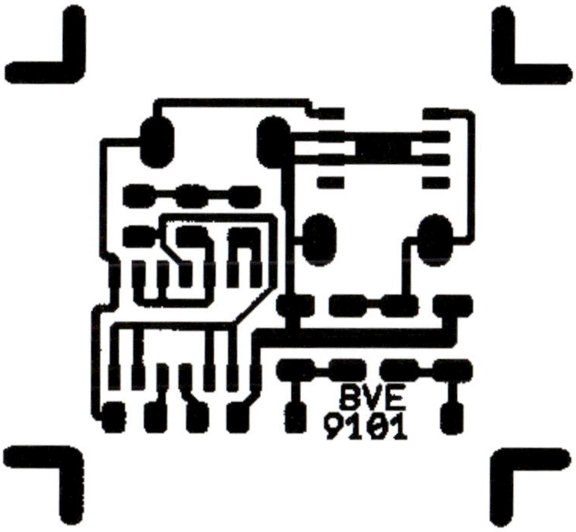

BVE 9101 component side (scale 2:1)

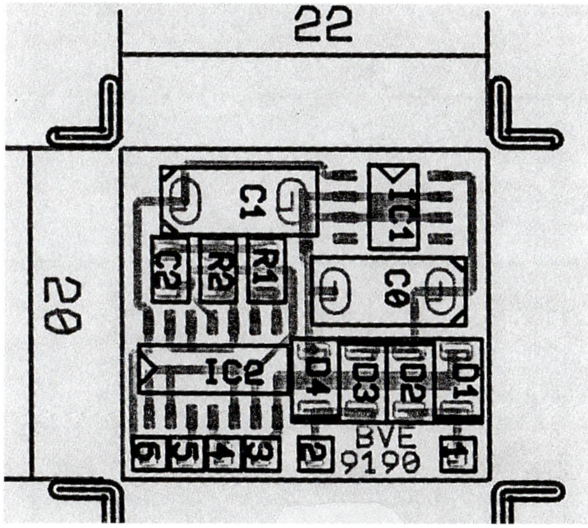

MINITIP

Address change on the Delta decoder

The Delta system provides you with the opportunity to run four engines with the central unit and engine selector (as well as the pilot engine address 80). There are four soldering tabs on the decoders for setting address codes.

It isn't at all necessary to buy an electric locomotive just because the position is unoccupied. You can run four beginner-set engines if you like. The only requirement is that the coding be redone. You'll need to solder a second bridge on the decoder. It is shown here. Adjoining the decoder illustration is information on how various solder bridge combinations provide different addresses.

Theoretically, four combination factors should provide 24=16 options. We have only used five. The mathematicians among you will have noticed that s eleven additional address combinations remain in the digital system. Others will be satisfied with the nine shown in the figure. In addition are address 54 with solder bridges for 1, 3 and 5, and the theoretical address where no bridges are used.

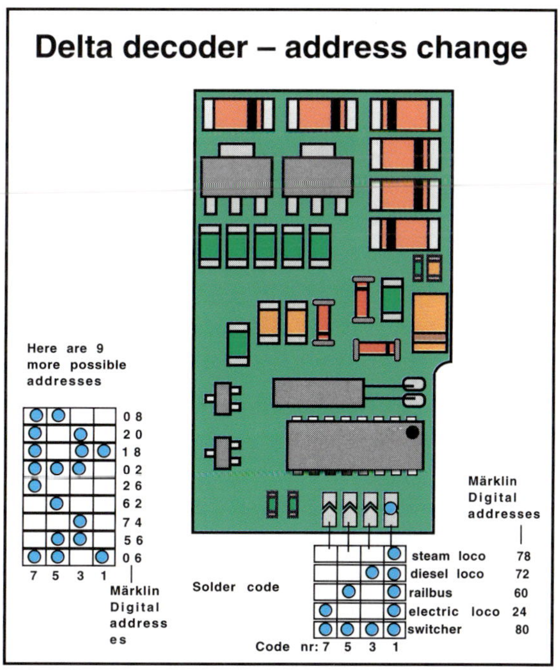

MINITIP

Pulsed Controller for HO

For common HO engines with rather robust motors, pulsed controllers are just the thing. As opposed to controllers with filtered DC to the rails – where voltage levels are increased or decreased with the controller knob – pulses are always sent to the trains at full voltage. At slow train speeds the pulses are short and spaced far apart. Longer pulses,at tighter intervals produce higher train speeds.

We often work with 60 or 120 Hz, depending upon if full- or half-wave rectification is used. This is because normal line voltage is 110 volts at 60 Hz. This frequency is ideal for model railroad applications.

But what should you do if you have battery power source or live in a country with a line voltage of 220 volts at 50 hertz? Although 50 pulses per second provides superb low speed operation, it is accompanied by an annoying hum from the rotor. Jumping up to 100 Hz removes the hum but also most of those sought after, low-speed characteristics.

This tip is a timer-controlled circuit that produces 60 pulses per second and can regulate locomotive speed, as well as set a maximum speed limit. The two transistors make up a so-called Darlington circuit – a circuit with an excellent amplification factor. T2 can manage loads with little heat loss. It's enough with a small heat sink. R7 is a PTC resistor with low "cold" resistance (0.6 Ω is common) making it suitable for circuit protection against overloads out on the layout.

Pulsed controller - components list

IC1	NE 556 double timers 555
R1	2.2 kΩ resistor
R2	100 Ω resistor
R3	1 MΩ potentiometer
R4	470 kΩ potentiometer
R5	15kΩ resistor
R6	270 Ω resistor
R7	PTC-resistor or 12V/24W bulb
C1	4700μF/25V electrolyte
C2	10μF / 16V electrolyte
C3,C4	20 nF capacitor
C5	47 nF capacitor
D1	bridge rectifier 50V/3 A
D2	diode 1N4001
T1	BC 140 or BC 637
T2	2N3055 power transistor
(1)	DPDT
(1)	Veroboard-experiments PCB
(1)	case

Pulsed train control with double timers

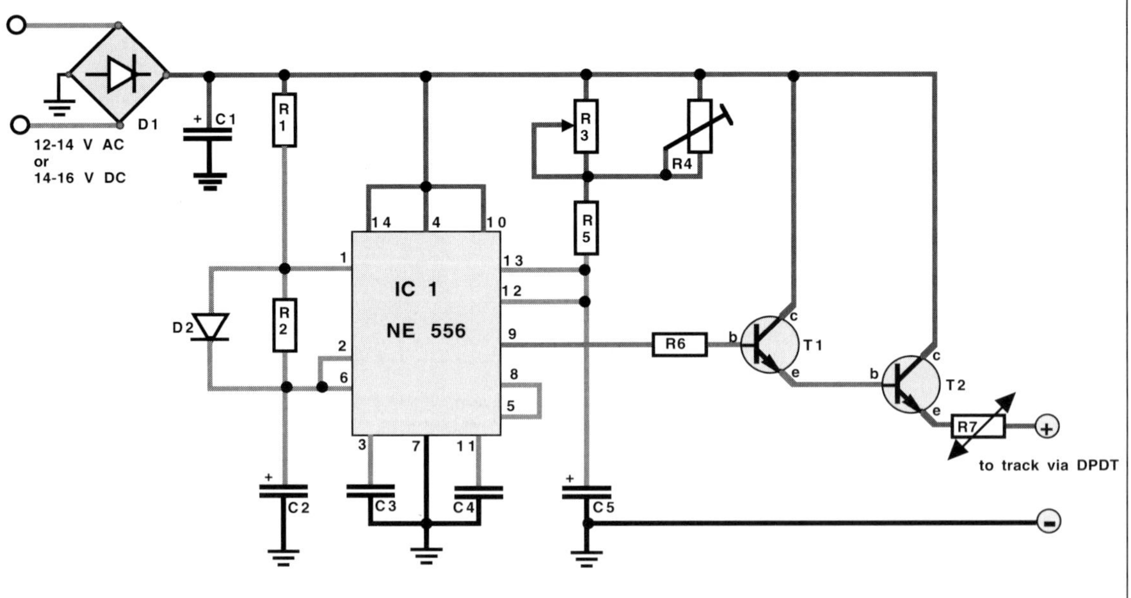

End-of-train lighting with quick-charging Gold Cap capacitors

Even though this end-of-train lighting circuit is both rare and relatively advanced for model railroad application, it's worth serious consideration. And I think that you'll fun with the technology behind it too.

Let's first take a look at the problem and then the solution. The last car on a train is always observed a little closer than others. And that's why we need end-of-train lighting that is both constant and stable. Constant lighting levels are difficult to achieve with analogue systems, with their varying running voltages to the trains.

Stable lighting is another problem in that power supply is not completely free from interruptions. Intermittent interruptions at block divisions and switches cause train lighting to blink. Power losses of longer durations shut down lighting completely. An example of this is when trains are stopped at passing sidings for a meet.

STABLE LIGHTING LEVELS DESPITE POWER GAPS

This project employs an uncommon capacitor that charges quickly. The capacitor is capable of powering end-of-train lighting without additional current for at least 30–60 seconds. I'll soon explain more about the capacitor, so let's start out with a look at the circuit as such.

The system surrounding the capacitor operates on approximately 5.5 volts. This is the highest permitted working voltage. The LEDs begin to glow at about two volts. In continual operation mode (continual charging), the LEDs will shine brightly, without interruptions caused by momentary power loss. When the train arrives at the station, lighting will remain on for a while after power is shut down.

The desired charging voltage level is obtained with resistor R1 and the Zener diode (when track voltages

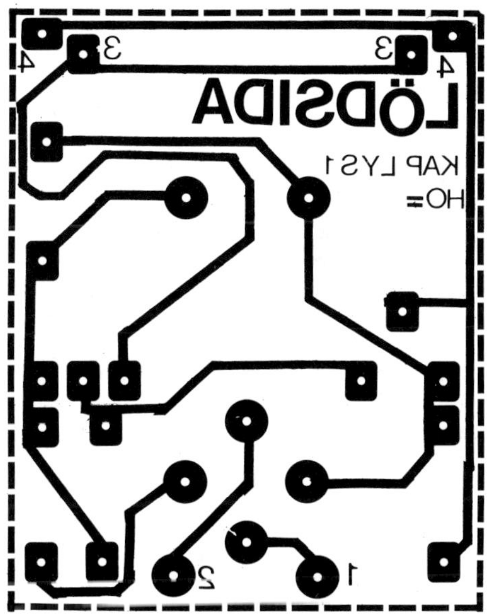

PCB (scale 2:1) Solder side

Quick-charging end-of-train lighting Components list

T1-T2	BC547 transistor
D1	B50C1000 bridge rectifier
D2	5.6 V Zener diode
D3-D4	3mm red diode (LED)
R1	390Ω resistor 1/4W
R2	10kΩ resistor 1/4W
R3	39Ω resistor 1/4W
C1	0.1 F/ 5.5V Gold Cap (supercap)

The electronics can easily be hidden in this gondola with a little "freight". In boxcars, even more space is available, allowing use of larger "gold caps". This will increase discharge time considerably with just a marginal increase in charging time.

exceed 5.6 volts). Discharge is through the constant-current circuit made up of the semi-conductors and the two resistors shown in the schematic.

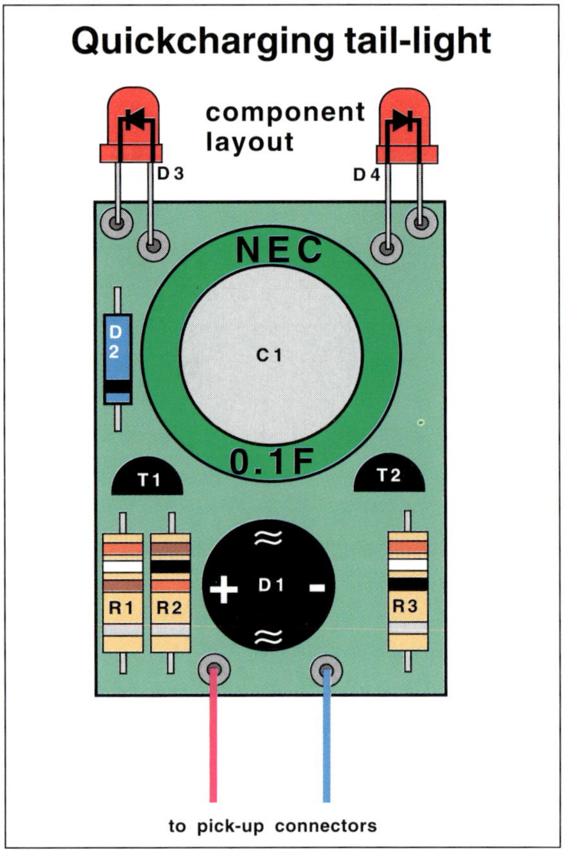

THE CAPACITOR IS LIKE A BATTERY

The Gold Cap (GC) is an unusual capacitor. Despite its small physical size, it has a capacity of 0.1 farad (or 100,000 μf). GCs are manufactured with 70 farad ratings and the usual operating voltage is 5.5 volts. These values make them ideal for computer RAM back-up applications.

The GC is a double-layered capacitor with a construction entirely different that other capacitors. Inside GCs are stacks of button-like cells. These normally have a working voltage of just under 3 volts. In the GC we will be using, there are two of these cells.

Within each cell are two layers of activated carbon and an aluminum overlay around the edges. The disks are impregnated with a dry electrolyte and contain no substances that are directly harmful. This is why Gold Caps are classified as environmentally friendly, in contrast to the most other components. The solvent used in the GC is propylene carbonate.

Because GCs have qualities that resemble those of rechargeable batteries, it is understandable why its environmental characteristics have been indicated. One of its "worse" competitors is the NiCd cell.

SEVERAL GOOD CHARACTERISTICS

A GC cannot be neither discharged to a level harmful to the component nor overcharged. Both of these are important qualities in our circuit Its operating temperature is in the -25–70°C range (-13–158°F). Furthermore, the capacitor can safely handle momentary voltage peaks (typically, increases from 3 volts to 7 volts).

Connecting them backwards is no real problem. Compare this to batteries that can explode when incorrectly connected or shorted. Actually, the cells as such are not polarity-dependent. It is only when they are connected serially that polarity must be determined.

Gold Caps are not suitable for filtering pulsed DC. This is because they have a much higher internal resistance than common electrolyte capacitors.

I hope that your curiosity has been aroused enough so that you begin incorporating Gold Cap capacitors in your own designs. Imagine using them as onboard power packs that would power locomotives through stretches of dirty track and over turnouts. That would be something!

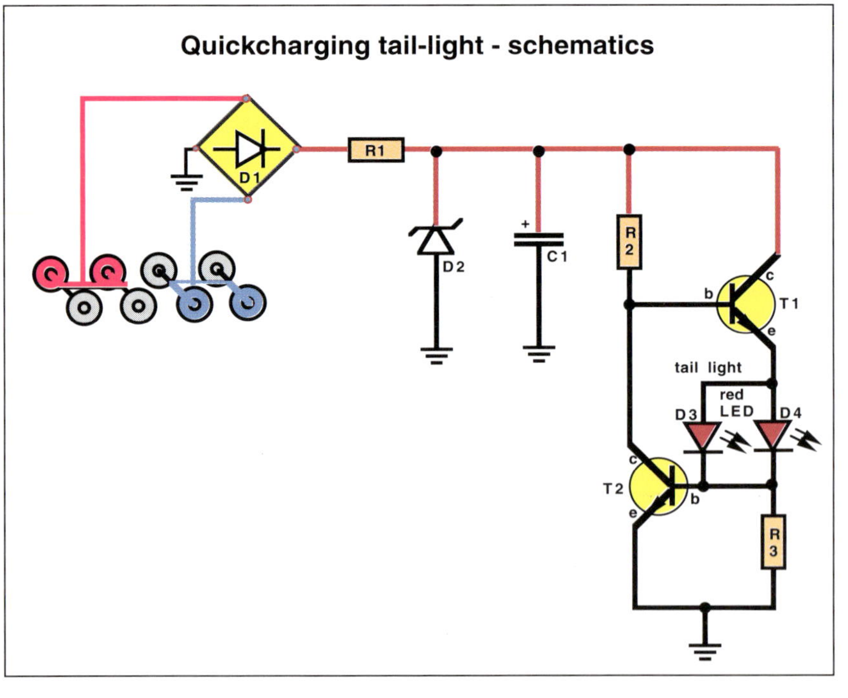

Adapter for bipolar switches

Although the bipolar electromagnet is a clever invention, it is often impractical for regulation from the switch control boards found on most model railroad layouts. The usual electric switches have magnetic coils with a common return lead in the middle. There are three connections. The iron core – that transfers power to the switch points – is drawn into the temporarily actuated magnetic field of one of the coils. It stays in that position until the other coil is actuated.

The bipolar electromagnet has just two leads. Connect plus to one and minus to the other to throw the switch. Reverse polarity to return the switch points to their original position. The iron core's direction of travel is related to the direction of current flow through the coil.

Several manufacturers offer bipolar switches. Among them are Lehmann Grossbahn (LGB) and Polk-Aristo. In its standard configuration, the control panel is set up for three-pole connection. One lead is always connected to the trackage, while the other two make momentary contact only.

Many applications employ bipolar electromagnets. On small, bistable relays, there is usually not space for two coils. Their many windings make them too bulky. Siemens, and some Japanese companies, currently use bipolar electromagnets on bipolar relays because they have but one winding.

ADAPTER FOR DIGITAL OPERATIONS

If a modern LGB turnout is to be regulated by Märklin's 2R system, an adapter is required between the bipolar switch and the digital decoder.

This adapter is so simple that you don't even need to make a special circuit board. You can quickly assemble your module with just four components and a piece of prefabricated circuit board. The component layout is shown on one of the figures. If you use the switch so much that the transistor overheats, add small heat sinks.

THIS IS HOW IT WORKS

Let's take a little look at its function – this can be helpful if should need to troubleshoot.

The electromagnet is activated by connecting the adapter's red lead to plus and the yellow to minus (color coding follows Märklin conventions). The voltage at the red lead raises the voltage at T2's base to a level higher than T2's emitter – an easy task, in that the emitter is connected

Adapter for bipolar electromagnet (LGB switches)

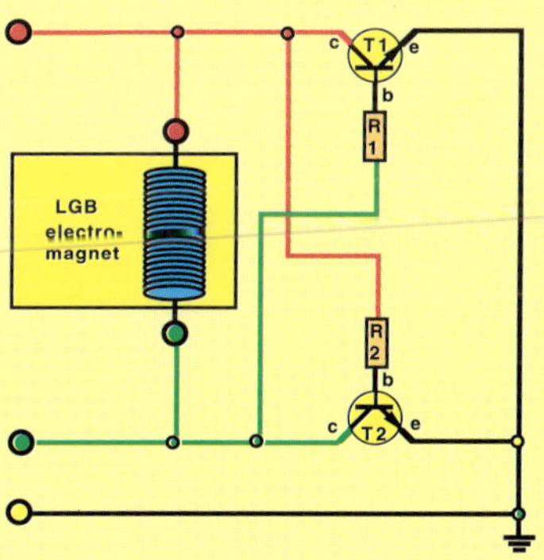

Adapter for bipolar switch machine - Components list

T1, T2	transistor TIP 41
R1,R2	resistor 4.7 kΩ, 1/4 watt
1 x	wire bridge
1 x	Veroboard card 40 x 40 mm

Layout on an experiment PCB

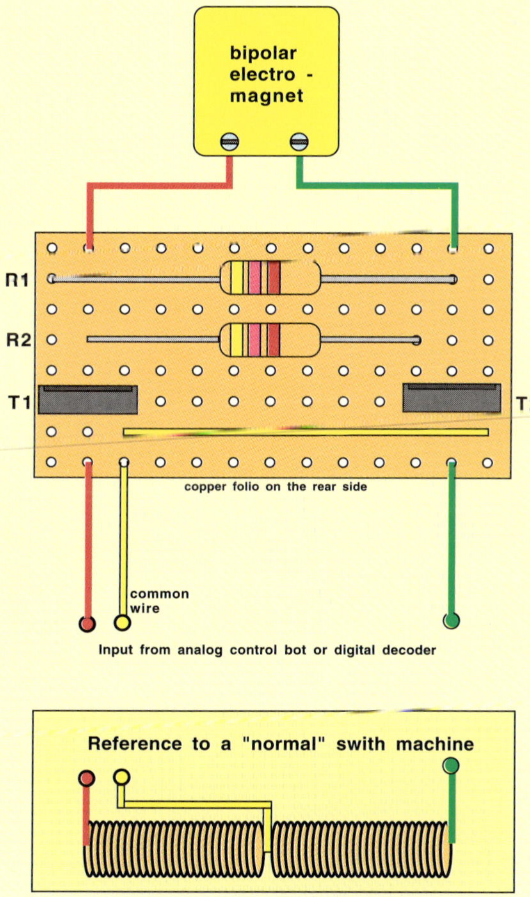

to ground. This turns on T2 and grounds its collector (at the green lead). The collector is also connected to T1's base via R1.

The ground potential ensures that T1 will not turn on and short the red connection to ground, thus defeating the purpose of the circuit. We have an additional lead connected to T2's collector that runs to one of the electromagnet's poles. To summarize up to this point The red lead to the electromagnet is now plus and the green lead is ground. This activates the coil.

AVOID SHORT CIRCUITS

To throw the switch, we now connect plus to the red connection. Initially, the red lead will be without voltage. Because T2's base no longer has a positive potential, T2 turns off. T1 turns on in that a positive potential – via R1 – is present at its base. The red lead is now grounded and the bipolar electromagnet is activated in the opposite direction.

If you have followed the above description, I am sure that you understand how the circuit works. While the plus voltage is being switched from the red to the green lead, and vice versa, there is an important point to keep in mind. Under no circumstances, may both red and green be connected to plus. This would create a short circuit.

MINITIP

The Delta central unit can easily serve as a booster

The Delta central is deceptive. It is much more than a simple beginners' toy for newcomers to command control. The unit is crammed with undocumented features. I've often wondered why Märklin hasn't promoted these features. The answer can only be related to marketing policy. Fortunately, we do not have to take such considerations. Just because you've outgrown the Delta central, and moved on to the Märklin Digital system, this doesn't mean that it can't continue serve a useful function on the layout. The Delta central can be easily employed as a booster – a device that transfers digital signals and, with an additional transformer, can power its own section of the layout.

Let's take a look at the figure. There are two minor, but very important, measures that must be taken. The first is that the two gray outputs must be bridged (use a short length of wire with two jacks). The second is that the controller handle must be in the far right, stop-position. It is important that the left side is deactivated. The blocks are divided at the third rail. Block division is achieved with paper insulation between the third rails of the two blocks, and with a triangular plastic shape that lifts the collection shoe over the block divider.

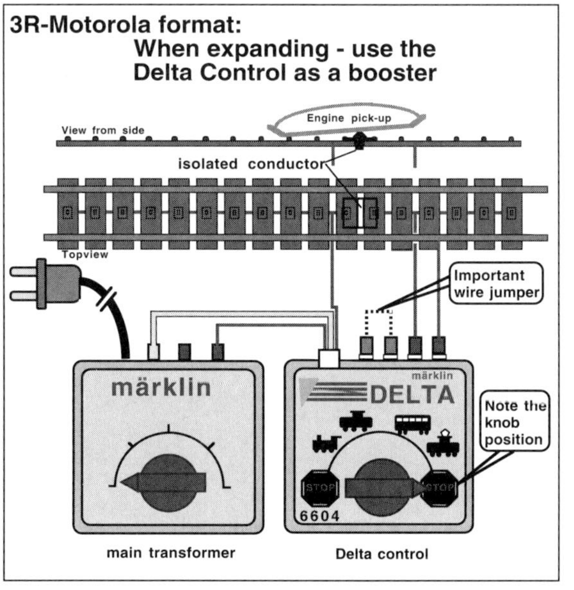

...and Delta as a 2R digital booster (DCC)

Note that the heading stipulates 2R and not 3R, a system also included in this section. The Delta central can under certain conditions, function as a booster. This is described in the related 3R minitip.

In line with what was mentioned previously, if we send a signal from the 2R digital system directly to the rails, the booster amplifies that signal and sends it on to the next control block. With an independent power supply, as shown in the figure, the booster can run a number of trains within the block. This is a less expensive alternative to buying a Digital booster.

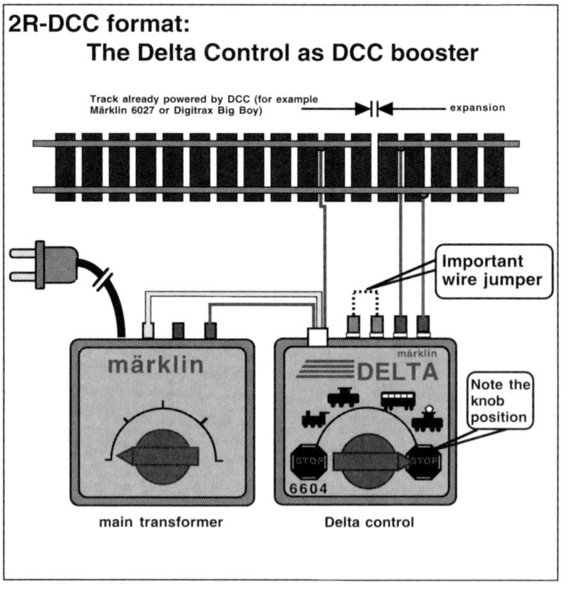

Dual braking for electric vehicles

We will now combine two different vehicle braking systems for Faller's Car System. This section is also about extremely small electronic building blocks – the kind built with surface mount components.

The more expensive Faller electric vehicles feature an electromagnetic braking system. The system incorporates electromagnets under the roadway and reed switches in the vehicles. There is a minitip to help you add this functionality to less expensive models.

If several cars run on the same loop – a common situation on heavily used roads – traffic jams can occur. Fast vehicles will tailgate slower ones. Intersections and grade crossings will also interrupt smooth traffic flow.

It is not practical to pack the underside of the roadway with stop coils, especially when jams can occur at any place on the road network. Before I describe the dual braking system, I should mention that you can set up control block divisions, exactly as on model railroad trackage. Only one vehicle at a time may run on a block. Vehicles approaching an occupied block must wait at a traffic signal – using a stop coil – before proceeding. The system works, but the resulting traffic flow is much more rigid than the prototype.

The author's test layout with IR controlled road traffic is seen in the foreground. A grade crossing is featured. When trains pass, road traffic is halted at the crossing with an IR transmitter. As new vehicles approach the crossing, they are brought to a smooth stoop, a half inch behind the vehicle ahead. At other locations (at bus stops, for example), braking is accomplished with electromagnets and reed switches mounted under the vehicles. This combination of two braking systems works well.

IR TECHNOLOGY

I've used infrared (IR) technology in the past (see the third book in the series) for train detection at grade crossings. It is both simple and reliable. We will now be using IR technology to detect approaching vehicles.

A one millimeter photo-transistor is mounted at the front of all vehicles. Photo-transistors are especially sensitive for IR light. This is the same technology as used on TV remote control units. The transistor should be mounted low, so that the vehicle's bumper serves as a visor. Direct the transistor so that it can detect IR light transmitted from vehicles ahead. Light from other sources can activate the detector, but with careful placement and a dark roadway surface, my own system has functioned well. The receiver features a sensitivity potentiometer so that the threshold between desired and undesired light can be set. Make adjustments during test running.

The entire receiver is mounted on a tiny circuit board. Its function is described in the accompanying figure captions. The board operates on 2.5 volts or more battery power. The motor is connected to the board.

As long as the unit does not detect IR light, it passes current to the vehicle's motor via a power transistor. As soon as the vehicle enters an IR field, current to the motor is dropped and the vehicle comes to a smooth stop.

IR TRANSMITTERS

At the rear of all vehicles are a small IR diode and transmitter. Its operating voltage range is at, and just above, 3.5 volts. This means that more power than normal is required. On the accompanying figure, I have shown how you can set up a power supply that can handle the increased load of the transmitter, receiver and motor combination.

The third battery can be installed on the bed of larger trucks and under the seats on buses. On smaller vehicles, such as fire engines, less space is available. You'll probably have to use the cab (and much patience) to install the third battery. Battery manufacturers offer a wide range of batteries in various shapes and sizes.

The transmitter is a separate unit. Its only function is to regulate the transmitter diode at the rear of the vehicle. It shares only the battery and main power switch with the other equipment.

Current is *pulsed* out to the IR diode (via a 555 IC) for three reasons. Firstly, we send relatively high currents through the diode to obtain a sufficiently broad spectrum of IR light – continuos current would quickly drain the batteries. Secondly, the diode would overheat with continuos current. And thirdly, by using high frequency pulses, we transmit an IR signal that differs from the ambient light. Although IR light is produced by common light bulbs, it is at a different frequency –60 Hz.

At the receiver input is a small and simple filter (performance is compromised somewhat by space requirements) that makes the receiver more sensitive to the transmitter's high frequency IR light than to other light. With the new technology that will probably be developed in the future, the receiver can made even more selective. The transmitter's high frequency can then be utilized even

The dual braking system

Schematics

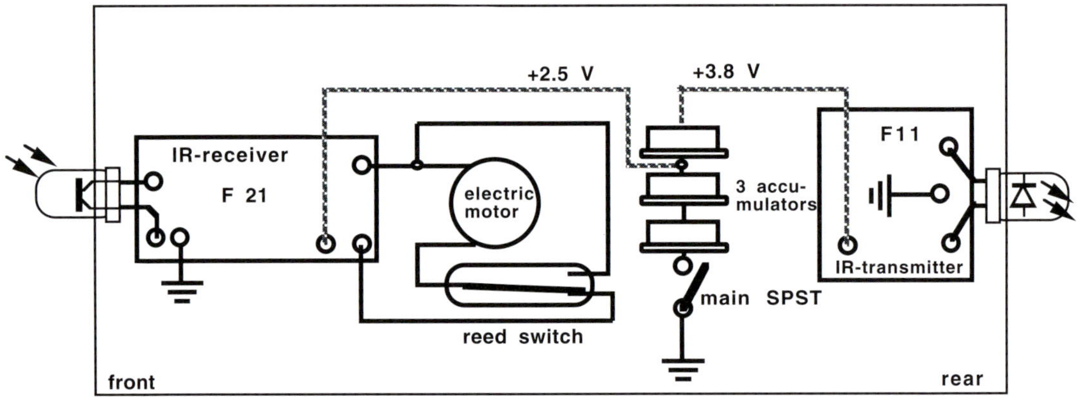

Here we have connected the various modules for two complete braking systems. To the left is a photo transistor placed in a car interior. To the right is a transmitter with its IR diode tucked away under the rear bumper. The reed switch is shown in the standby state which means normal operation for the motor. Note that the motor is intentionally short circuited in the reed switch's activated position (in a magnetic field) so that the car stops quickly.

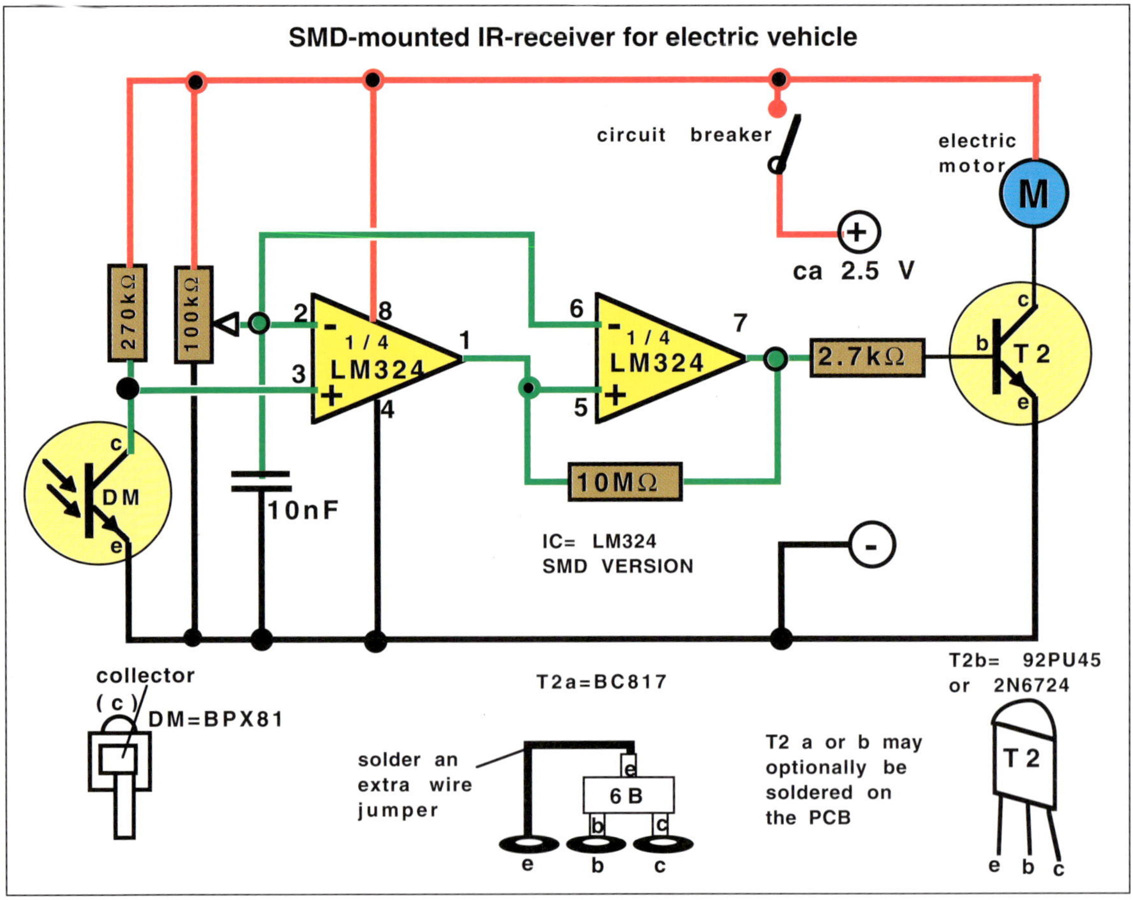

Dual braking system as on the prototype

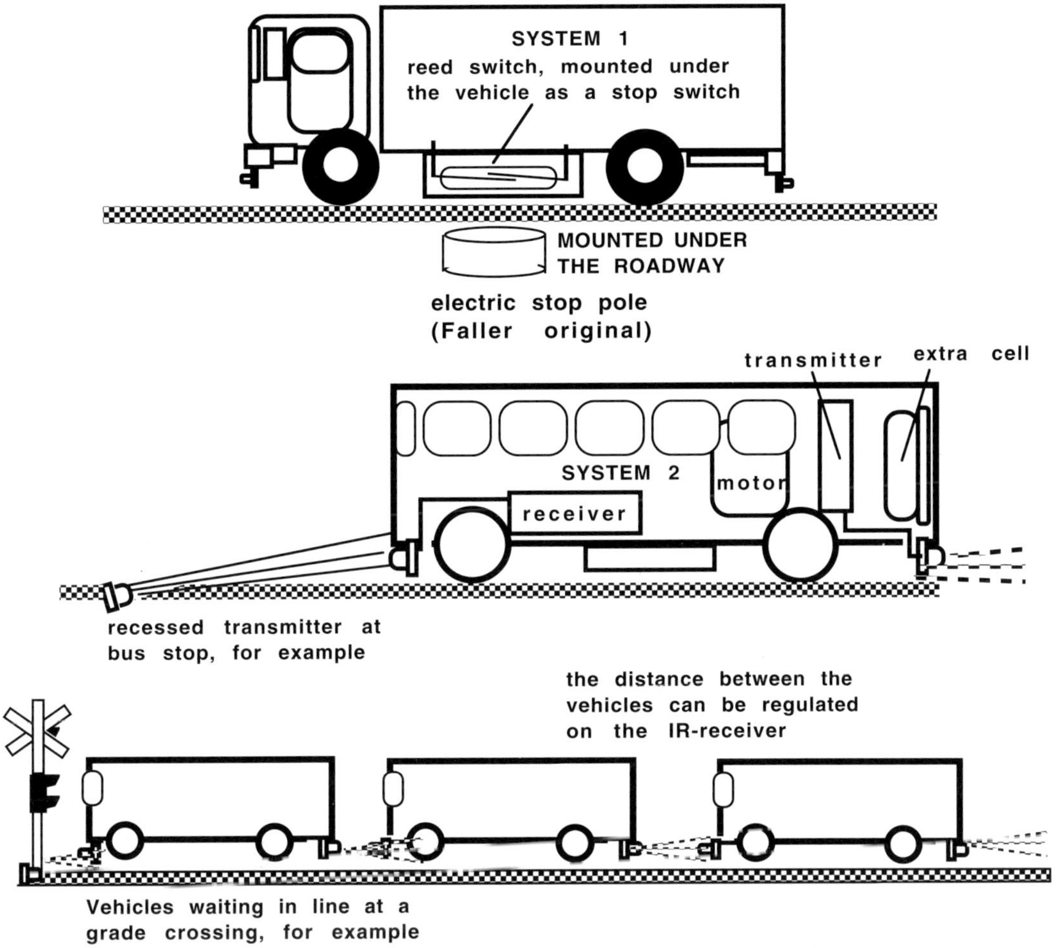

Above: Here is a system diagram showing combinations of the two braking systems. Vehicles waiting in line would not be possible with Faller's standard braking system. On the other hand, the magnet controlled system is foolproof for stopping vehicles with low battery voltage levels.

Right: The IR transmitter requires a higher voltage than the motor. This means that one to two extra cells must be mounted on the truck bed. This isn't usually a problem. Buses offer even more space.
By pulsing current to the IR diode via the 555 timer, battery power is conserved.

Left: The receiver consists of four electronic blocks. To the left is the IR transistor that leads current to ground when activated. The level at pin 3 is altered depending upon if IR light strikes the transistor or not. The second part is a comparator – the trimpot determines the threshold – with the output at pin 1 on IC1.
The other operations amp increases signal strength and the last block is the power transistor that switches motor power on and off.

Schematics for SMD-transmitter

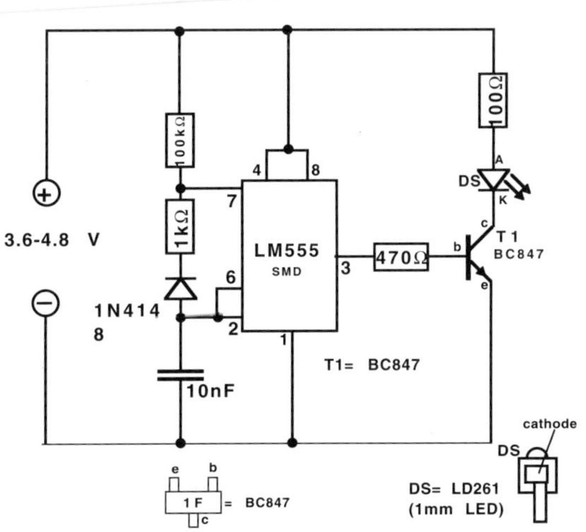

PCB
(scale 2:1)

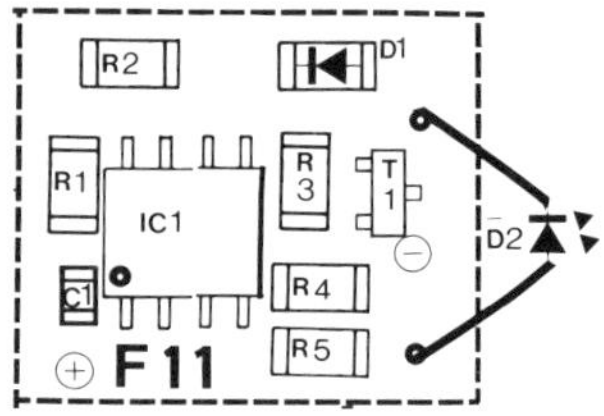

Component placement for F11 is relatively easy but requires a steady hand. There is a special section in this book devoted to assembly techniques. Be careful that excessive solder is not used at the copper lead between IC 1's pins 3 and 4.

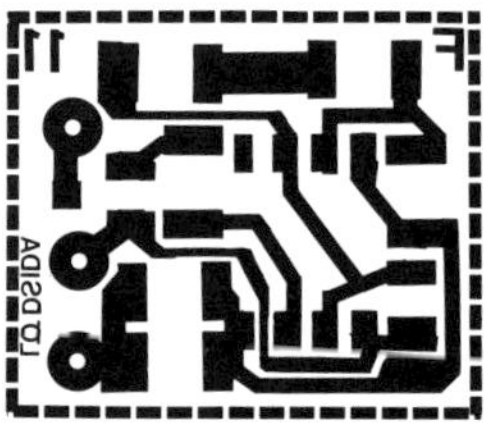

The circuit board for the transmitter is small – measuring just 1/2" x 7/16". Holes need only be drilled for connection leads in that components are mounted on the board's surface. Four 0.6 mm holes are sufficient. Single conductor wire normally used for windings can be used. It is available in several colors.

Components list for SMD-transmitter

IC1	**LM555**
R1	**100kΩ**
R2	**220kΩ**
R3	**470Ω**
R4,R5	**100Ω**
C1	**10nF**
T1	**BC 847**
D1	**1N4148 (SMD)**
D2	**LD 261**
1 x	**circuit board F11**

Mounting of components should begin with the centrally placed IC, followed by the trimpot and resistors. Although the techniques used in soldering are found in a separate section, a few words of adivice are in order. The trim pot has soldering points on its underside only. Melt a small amount of solder to these surfaces and those on the board where the trimpot is to be installed. With the component pressed against the board, apply heat with the soldering pencil at the mating surfaces until the trimpot sinks into palace.

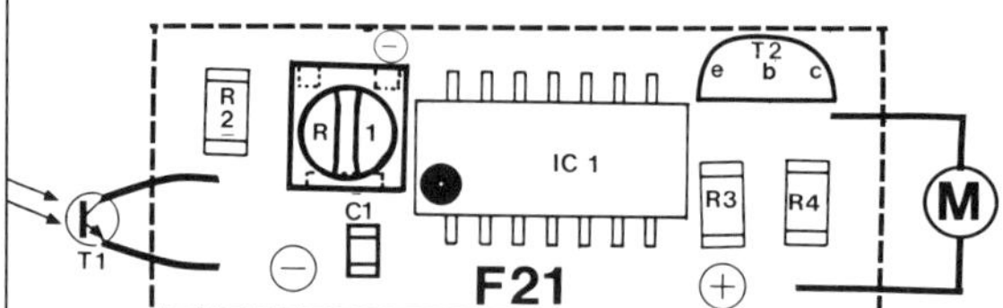

Components list for SMD-receiver

IC1	**LM 324**
C1	**10nF**
R1	**100kΩ SMD trimpot**
R2	**270kΩ**
R3	**10 MΩ**
R4	**2.7kΩ**
T1	**BPX 81**
T2	**BC817 (see figure)**

The circuit board (in 2:1 scale) for the receiver is long and narrow to fit into Faller's newer miniature vehicles. The correct dimensions are 26 x 10 mm. Seven 0.6–0.7 holes are normally drilled if the larger transistor types is used.

In smaller vehicles, surface mounted transistors must be used and of the four hole markings at the corner of the board, only the two inter-connected outer ones need be drilled. The other two serve instead as soldering points for the transistor's base and collector leads.

better than what is possible at present.

The transmitter's range to the rear is 10–18 centimeters (about 4–7 inches) – varying with available power. Within this range, approaching vehicles brake to a stop. This works well. The only place where I have had problems is in an especially tight curve, where the photo transistor's "angle of view" is too narrow to detect the transmitted IR light. I'll most likely remedy this problem by straightening out the curve a bit and continuing my experimentation with dual transistors (connected in parallel and slightly angled away from each other). I am also expecting new photo transistors on the market, with a wider reception angle.

SURFACE MOUNTED COMPONENTS

Faller's electric vehicles are miniaturized, technical masterpieces with a high degree of reliability. Ten years ago, they would have been impossible to produce – the required technology did not yet exist. We are dealing with small, self-powered units in HO scale. To find room for all the extra gear that I have described here, we must use the most compact electronic construction technology available, namely surface mounting.

Don't worry. This technology is not something reserved for specialists in white coats, equipped with specialized subminiature tools and electron microscopes. I have none of this equipment and am neither a specialist nor skilled technician in surface mounting. I am a hobbyist armed with a common soldering pencil fitted with a fine tip and am also somewhat near-sighted. That's all. The only difference between our constructions and those assembled by the professionals is that ours will take more time to build and the components will probably be mounted a bit askew. In spite of this, our results should be nearly as good as the pro's. As far as time goes, I hope you have enough to spare for hobby activities.

There is a surface mounting section earlier in the book that covers this subject in great detail. Read it before you begin building these IR modules. There may even be something there for those of you that have worked with surface mounted components before.

MINITIP

Controlling locomotive lighting with Delta

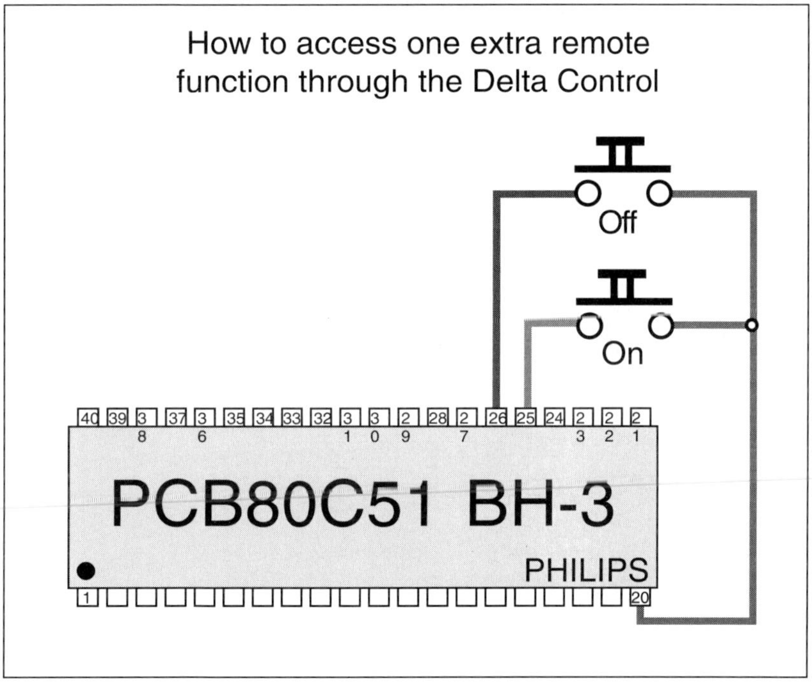

To motivate the difference in price between the Delta and Digital systems, Delta offers fewer features. Let's use locomotive headlights as an example. A Delta locomotive's headlight is always on. There is no function for switching the headlight off. Light intensity is constant but only if the locomotive is running on digital trackage. If the engine is run over analog trackage, light intensity is proportional to track voltage.

I have previously covered the decoder modifications that make possible headlight control with the Digital central unit. We will now add this functionality to the Delta central unit.

WARRANTY VOIDED

To make the needed changes, you will have to open the unit's case. This voids the warranty. On the brighter side, the procedure is easy to perform. Just drill away a few rivets, lift off the controller handle and separate the casing to reveal the unit's circuit board, equipped with one large 40-pin IC.

The figure shows how to connect two closing push-button switches. For the more experienced, a good solution is to solder the leads directly to the pins (as shown in the figure) and drill holes in the blue case to accept the buttons.

When you are finished, you can control lighting or any other accessory connected to the F0 function.

Component descriptions

This section is intended for reference use. You can't keep all pin connection and electrical data in your head. Now and then, you must use reference material. And when you need help, it's good to have reference close at hand.

This collection is by no means complete and it doesn't offer detailed information on the listed components either. Each IC is usually documented with five to ten pages of reference material from the manufacturer. Many discrete semi-conductors are also extensively documented. Your supplier normally has access to this material and can probably supply data sheets if needed.

Instead of depth, I have concentrated on breadth. A broad range of facts, color code information, etc. are given for the majority of components used in this book. In some cases, related components have been included for situations when circuit modification is desired.

There is also a symbol chart showing the most common electrical components and how they are indicated on electrical schematics.

LTS10304E
Jumbo 7-segment display

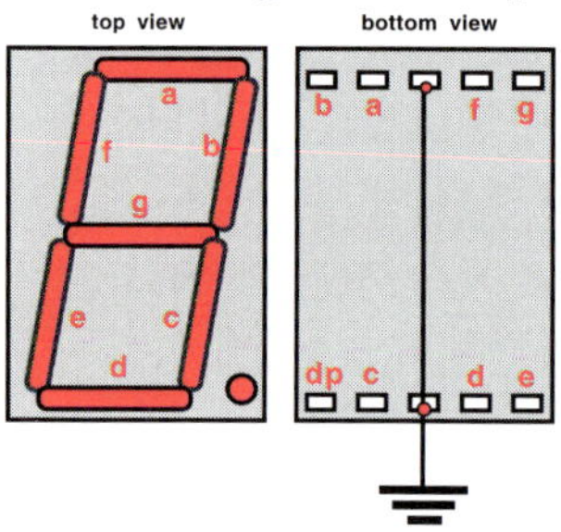

Color code for Tantal Caps

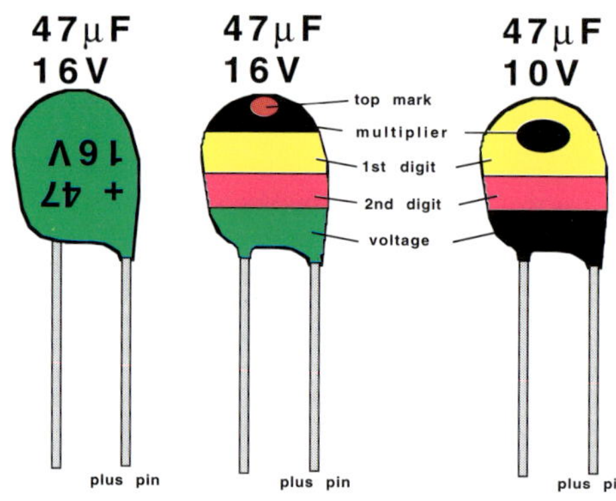

color	black	brown	red	orange	yellow	green	blue	pink	grey	white
digit	0	1	2	3	4	5	6	7	8	9
multiplier	x1µF	x10µF	x100µF	—	—	—	—	—	x0.01µF	x0.1µF
voltage	10V	—	—	—	6.3V	16V	20V	—	25V	3V

Color code for Caps

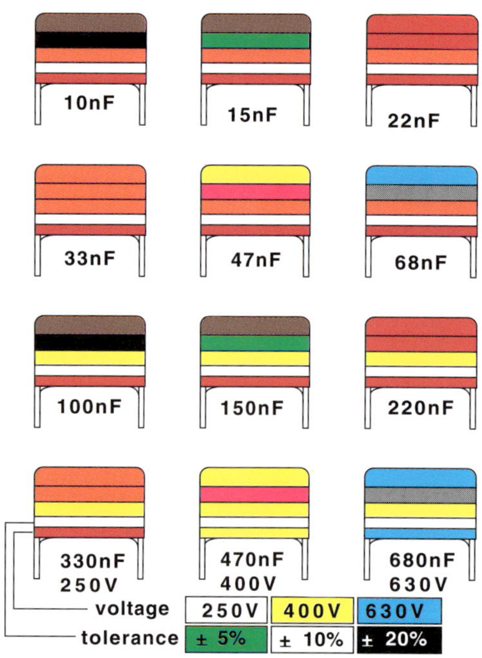

Code for ceramic caps

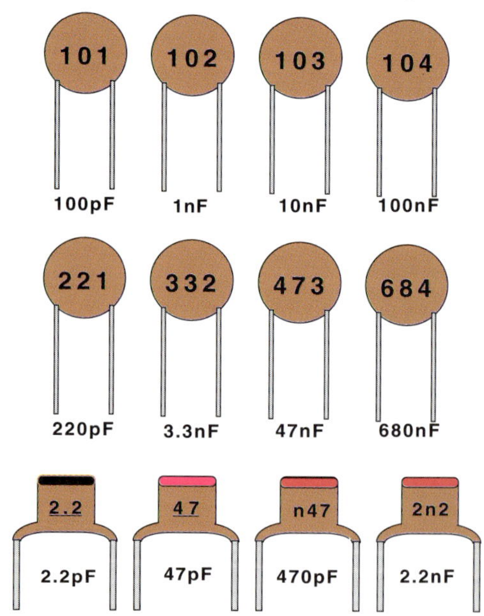

It should, however, be noted that certain symbols used in this book do not conform to those traditionally used. This was for practical reasons. Using the alternative symbols permitted me to prepare the schematics on my computer. Hopefully, this will not be a problem.

The older symbols for logic gates have been used. These too, are better suited to computer assisted drawing. For Boolean algebra, as applied to gate logic, see the third book in this series.

How to translate different color codes

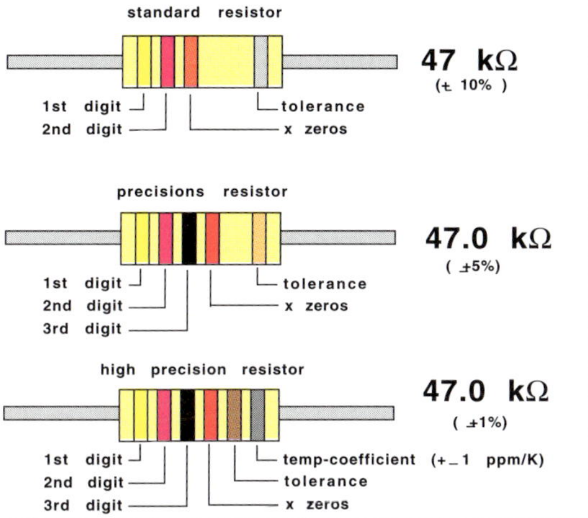

Opto-components

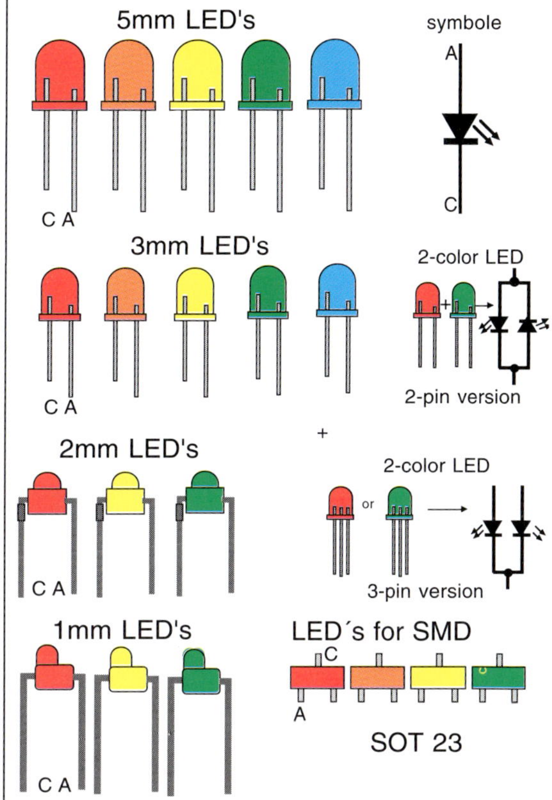

Resistor color code

color	digit	x zeros		tolerance %
black	0	10^0	1	20
brown	1	10^1	10	1
red	2	10^2	100	2
orange	3	10^3	1000	3
yellow	4	10^4	10000	100
green	5	10^5	100000	0.5
blue	6	10^6	1000000	0.25
purple	7	10^7	10000000	0.1
grey	8	10^{-2}	0.01	-
white	9	10^{-1}	0.1	-
gold	-	10^{-1}	0.1	5
silver	-	10^{-2}	0.01	10

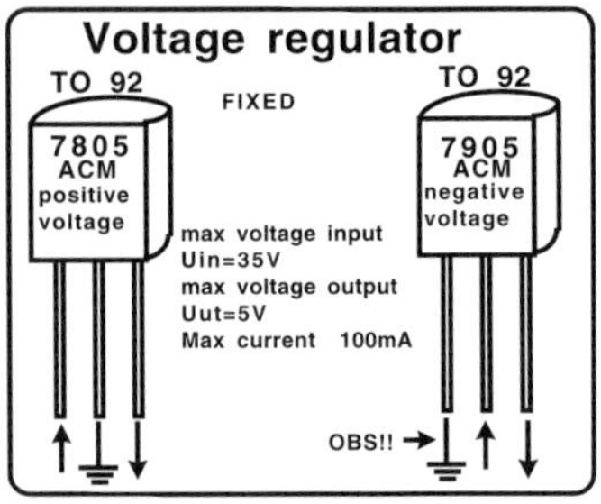

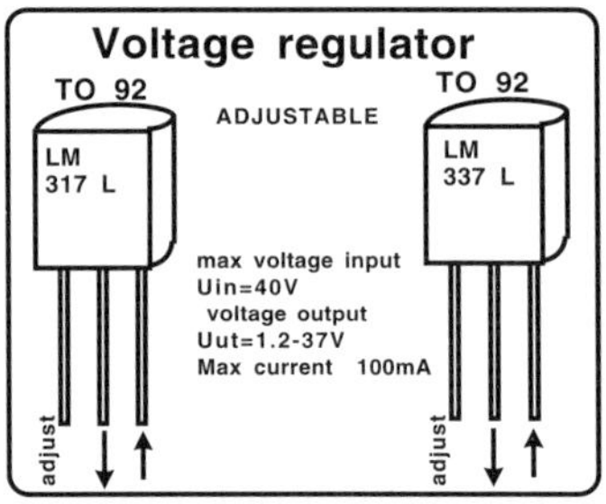

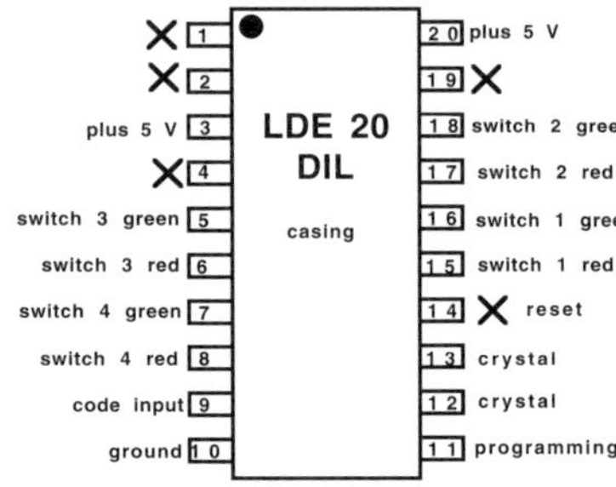

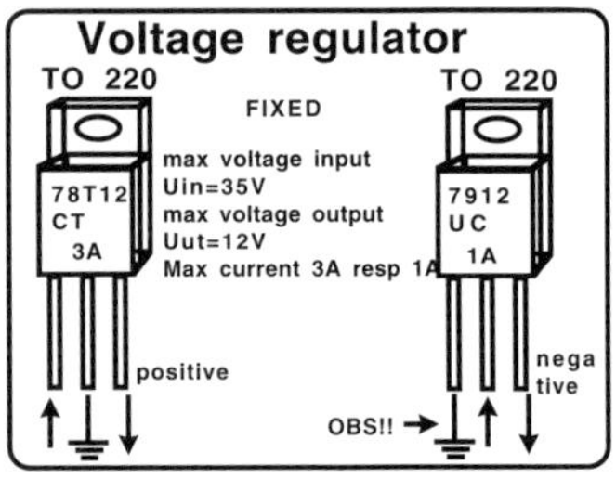

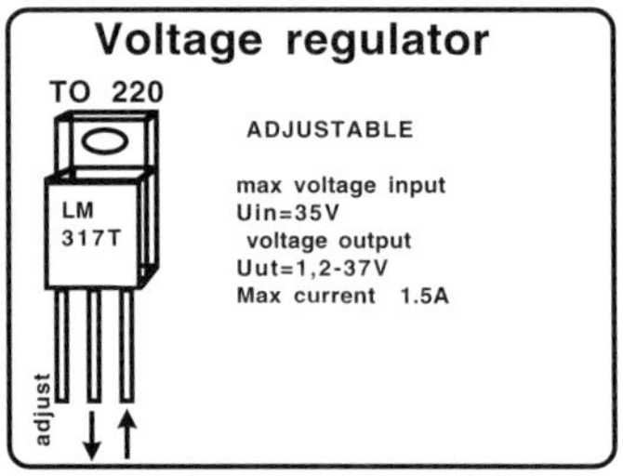

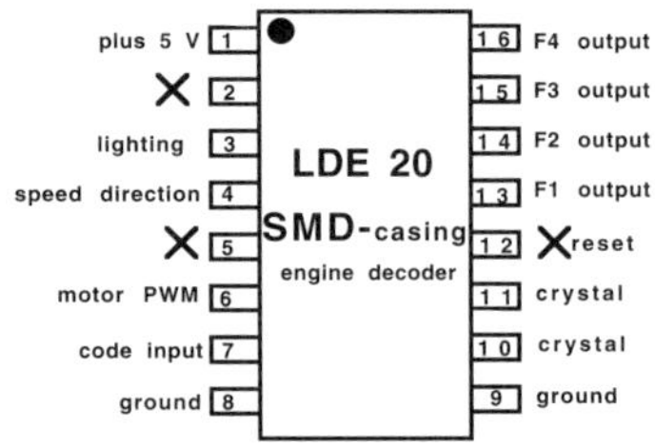

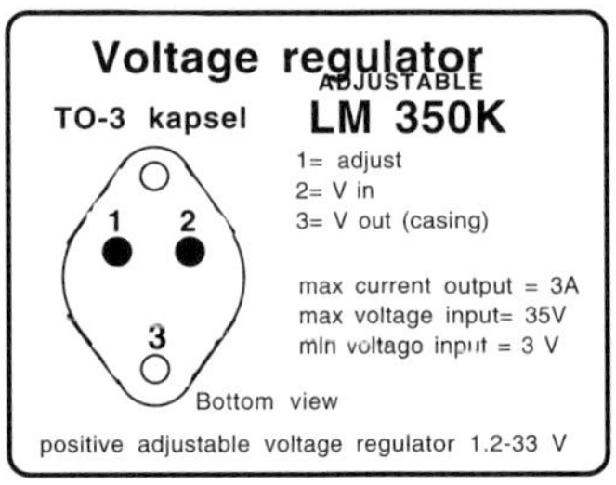

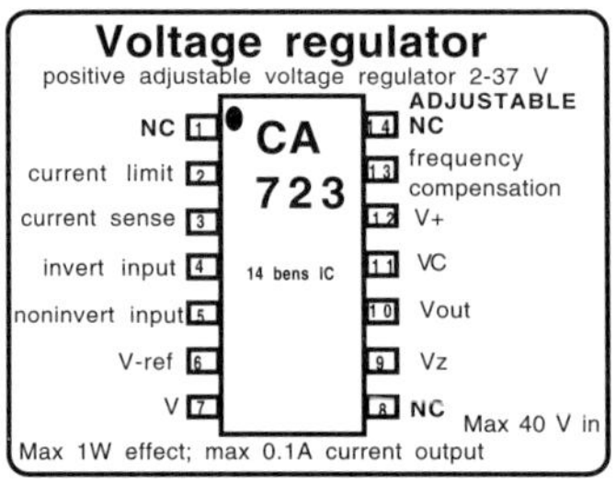

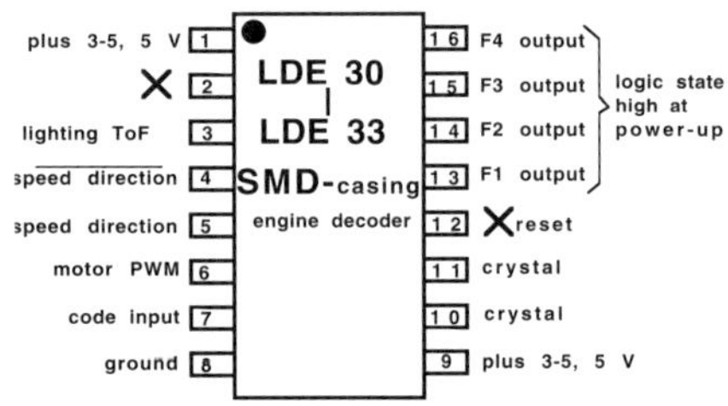

SMD-Semiconductors

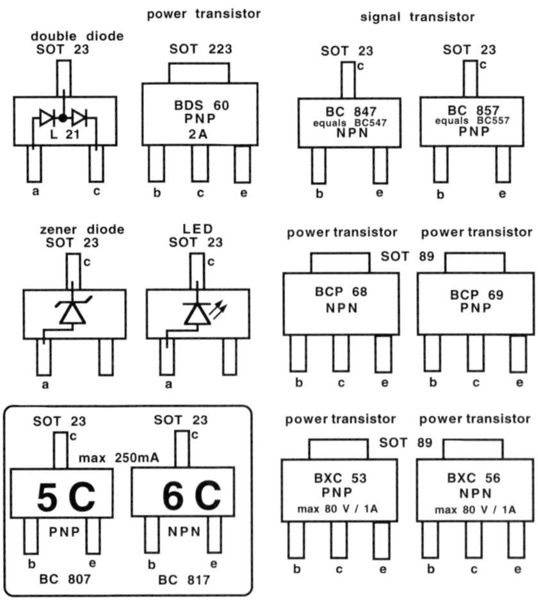

Symboles

logic gates and op. amps.

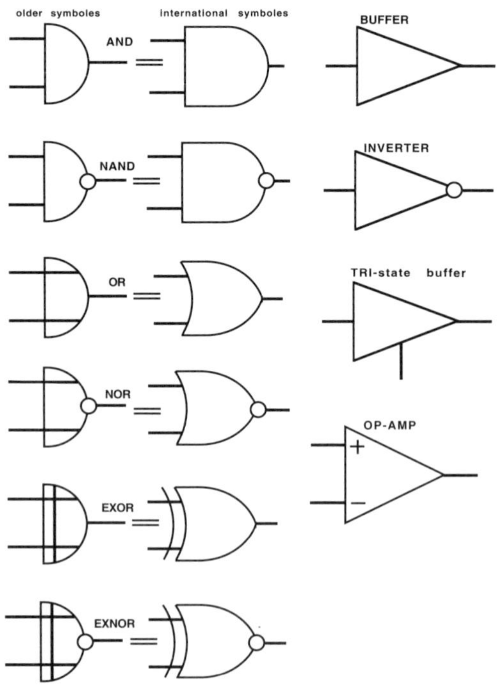

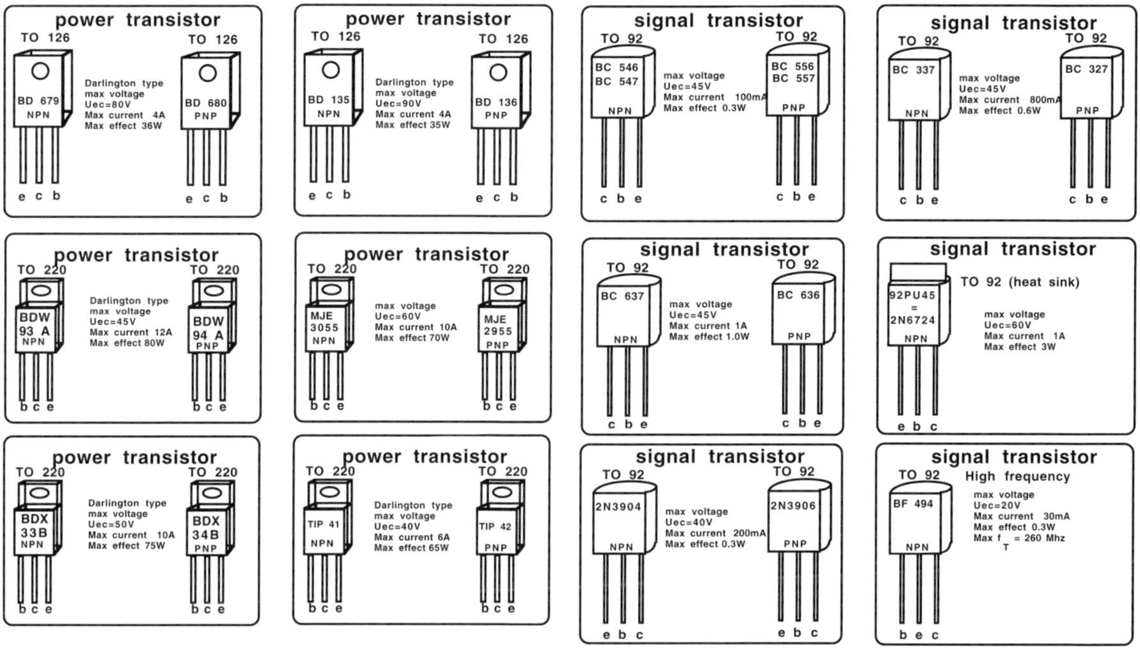

Electric Symboles

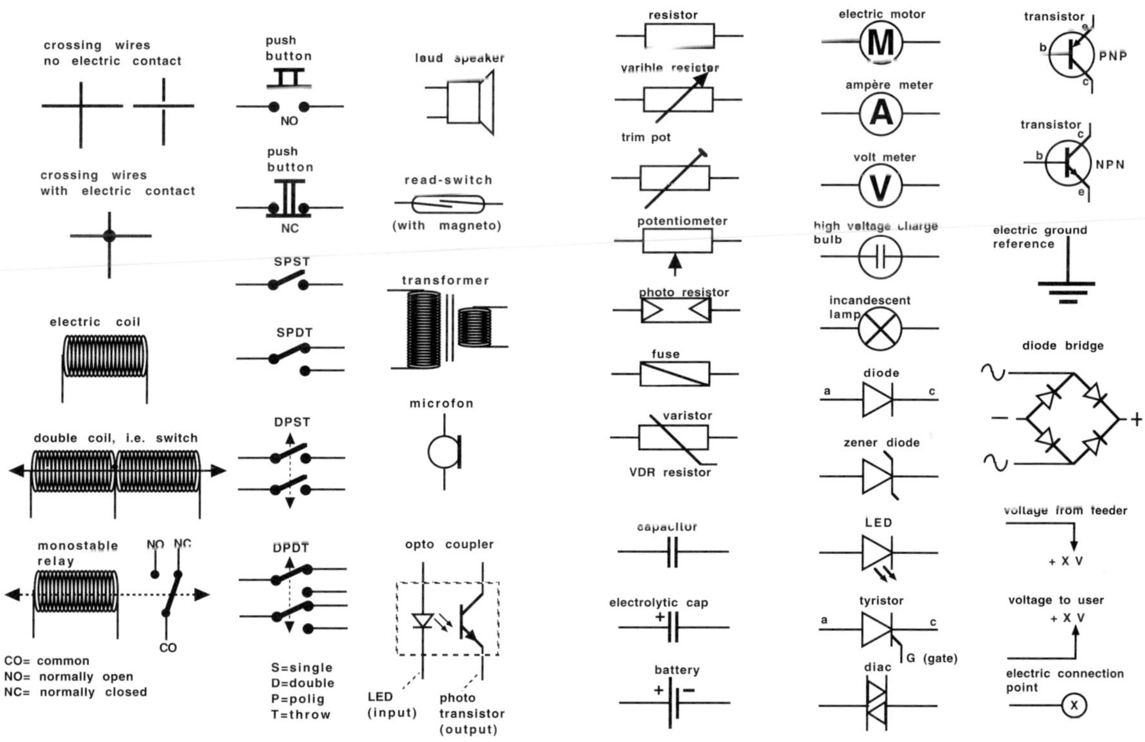

Bulb guide

Bulbs are available in a wide range of sizes and configurations. With respect to the various construction projects in the this book series, we will concentrated on the smaller bulbs. See also the information on light emitting diodes in the component section.

Bulbs with wire leads are meant for soldering to another wire or contact point for power supply. This type of soldering should only be done using a soldering pencil or small soldering iron. Note that the leads are not insulated. To reduce the risk of short circuits, use insulating paint or a short length of insulation from another suitable wire, trimmed to the correct length.

There are many sources for these type of bulbs. See the book's introductory section for information on suppliers.

Small bulbs for E5.5 sockets

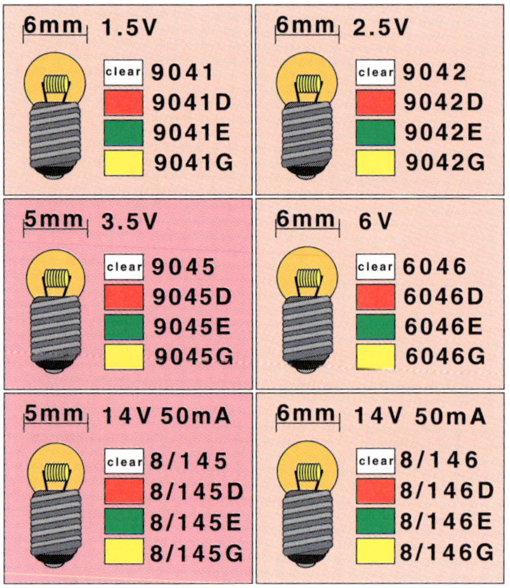

Small bulbs for wire lead connections

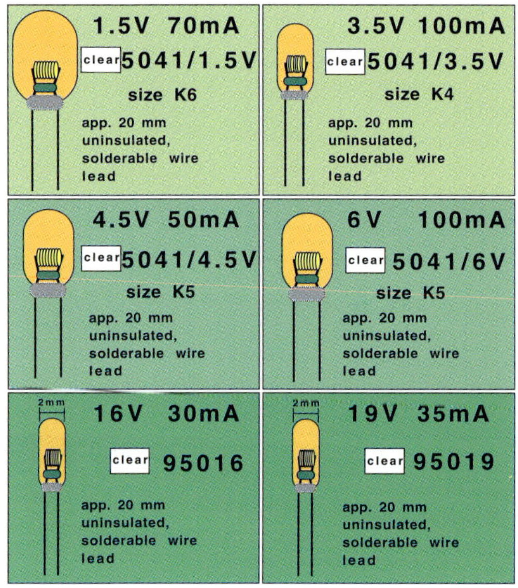

Small bulbs for E5.5 sockets

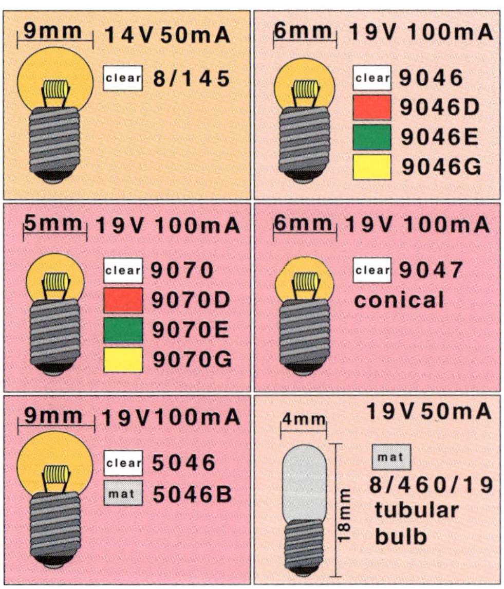

Small bulbs for bayonet sockets

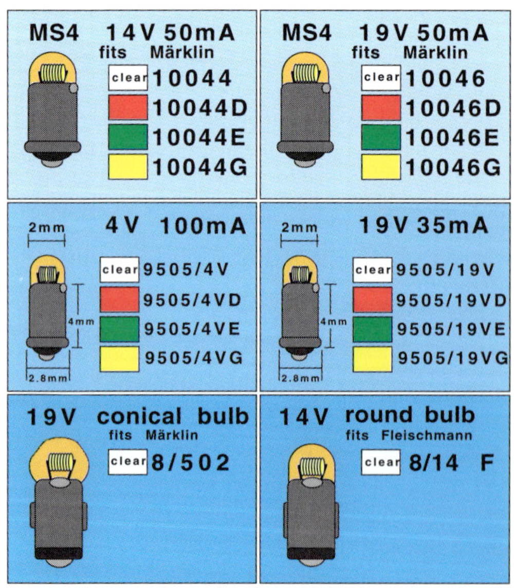

MINITIP

How to make your own Delta stationary controller

When you run trains with the Delta system, the controller handle on the power pack is used to regulate train speed and direction. Locomotive selection is made with the Delta central unit's selector handle.

There are two reasons for running trains with a controller other than the one incorporated in the power pack. The first is that slow starts are difficult to achieve with the built-in controller. This is because the controller's minimum voltage is greater than what Delta-equipped locomotives require. And when the engine is running, the lowest atttainable speed is much too high.

Secondly, you may want to reserve a section of the layout for analog operations. Perhaps you have older engines without decoders and you would like to keep them in service. It is then good to run a section of the layout with the power pack's analog controller. Connect it as usual between the red and brown contacts.

You will now need a second controller for Delta engines – one that provides slower starts. The figure shows a simple circuit that is connected between the power pack and the Delta central unit.

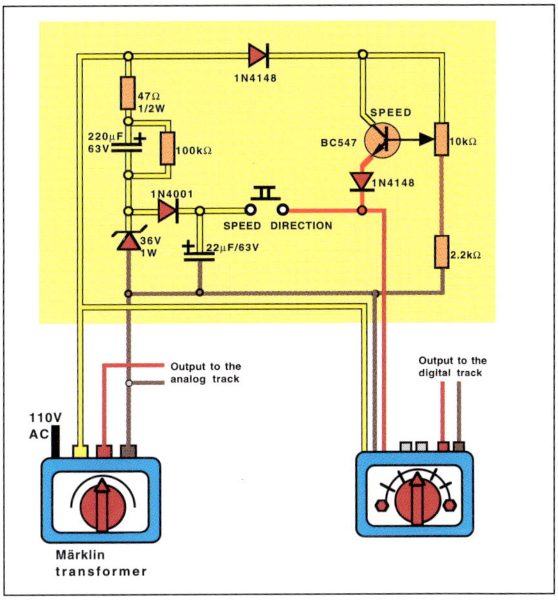

Train speed is regulated with a 10 kΩ potentiometer with a closing pushbutton. Mount the unit in a suitable enclosure. The Delta central unit will continue to function as usual.

MINITIP

How to make your own Delta handheld controller

The Delta central unit features a locomotive selector. It is used to select one of four possible engines for regulation. Speed is controlled by the controller handle on the power pack. While it is possible to run four engines at once, only one may be regulated.

If you look at the rear of the Delta central unit, in addition to the red and brown track connection terminals, are two gray female terminals. You can use these to connect a handheld throttle. This will add the capability to run a fifth locomotive. The fifth locomotive may be run independently of the other four.

The circuit described here is less expensive, more interesting and offers slower running characteristics than available commercial throttle's.

The figure shows how the unit is built using just three electrical components. Use a small enclosure to house the circuit with a long (10–15 feet) two-wire cable for mobility.

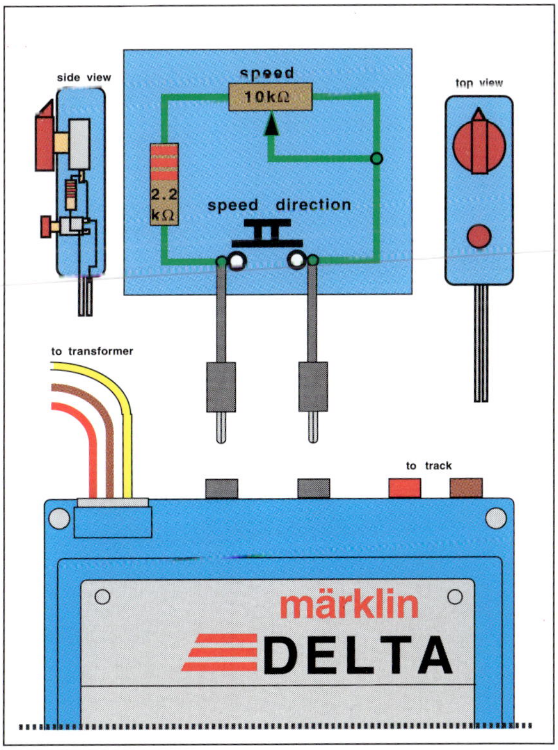

Special ICs

4013

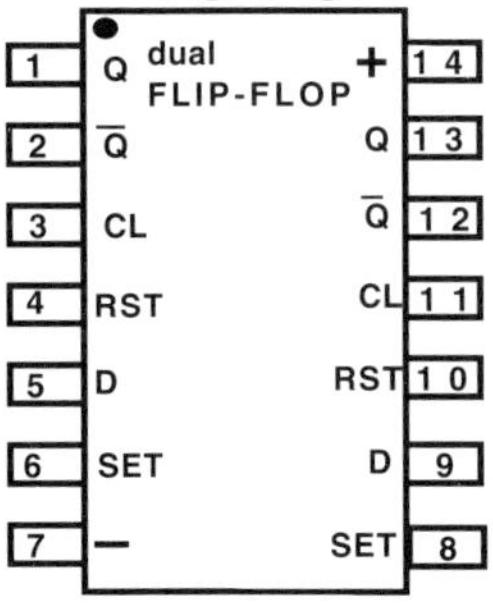

Q and Q̄ are outputs and are always complementary, i.e., they have opposite ligic levels. D is the data input and CL is the clock input. SET 1 sets output Q and RST resets ouput Q. Pins 1-6 constitute one circuit and 8-13, the other.

4017

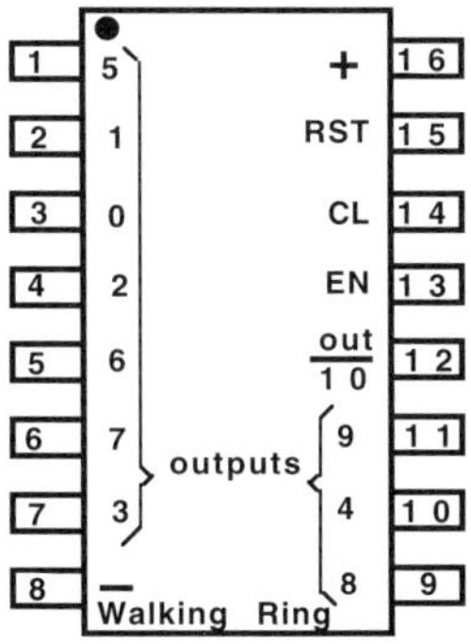

This counter divides by 10 and activates one of the 10 outputs. It counts in assending order and restarts at 0 – that's why it's called a walking ring. Normally, EN and RST should be 0. On step is made for each positive pulse at CL. At pin 12, an output pulse comes every 10th pulse.

4020

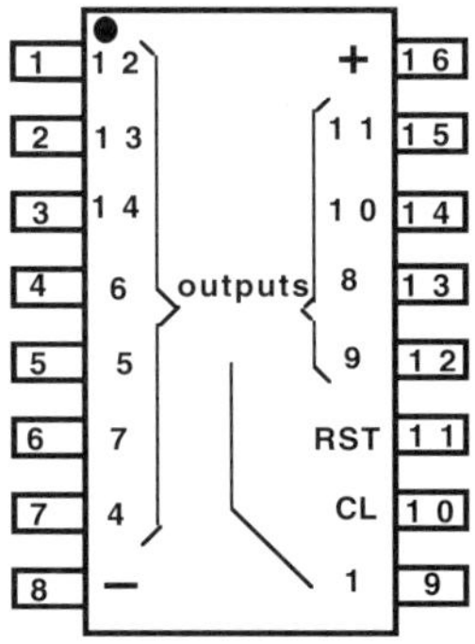

This is a 14-step counter that can divide an input pulse by as much as 16,384. It is refered to as a ripple counter in that it reacts to the change phase of the logic state at the input, and changes the outputs in sequence. It counts binarily and lacks outputs for the 2nd and 3rd steps.

NE-555

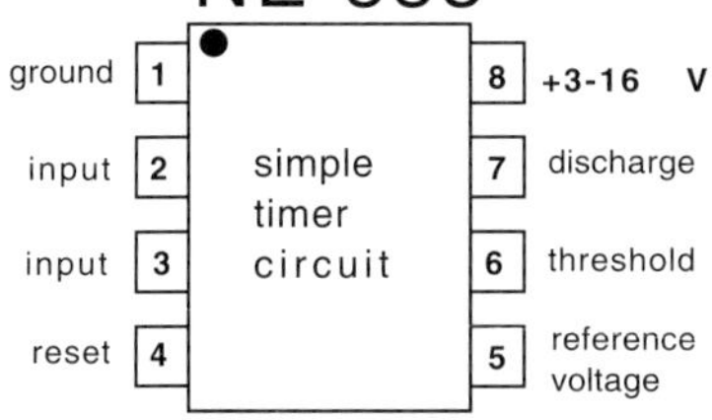

There is a special CMOS variant of the NE 555 designated 7555 that use just one third of the current required by the 555. It is suitable for small battery circuits. The highest permited output effect is 0.6 watt.

NE-556

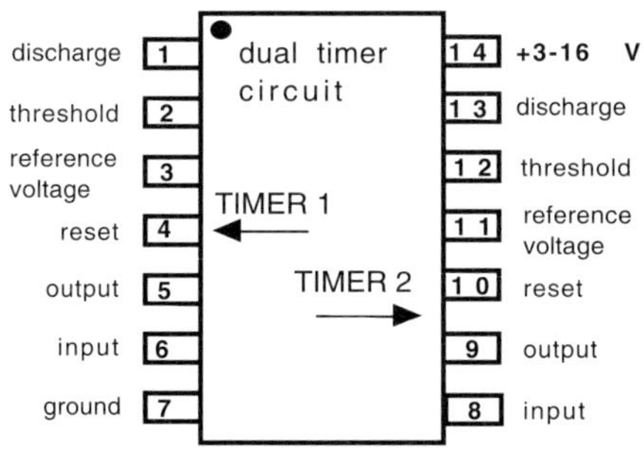

Two 555 timers in one case.

L 200

Case: 5-pin PENTAWATT

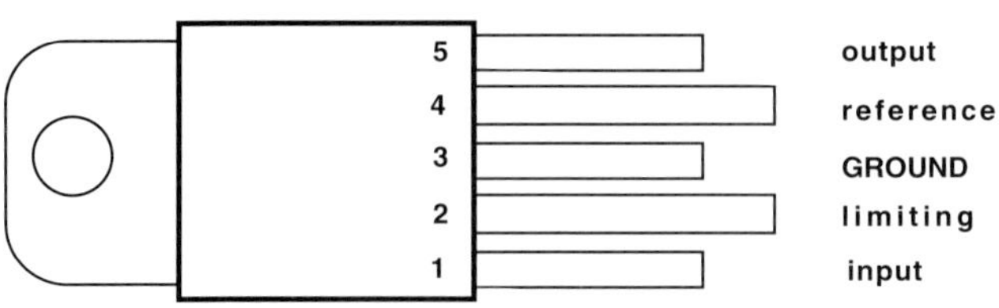

Max input voltage: 40 V; max output current: 2 A
Output voltage variable from 2.85 to 36 V

Metric system

Units of length
1 m = 10 dm
1 dm = 10 cm
1 cm = 10 mm
1 cm = 0.3937 inch
1 inch = 2.54 cm

Units of weight
1 kg = 1000 g
1 ounce = 28.35 g
1 pound = 0.4536 kg

CMOS cicuits with buffer gates

4049

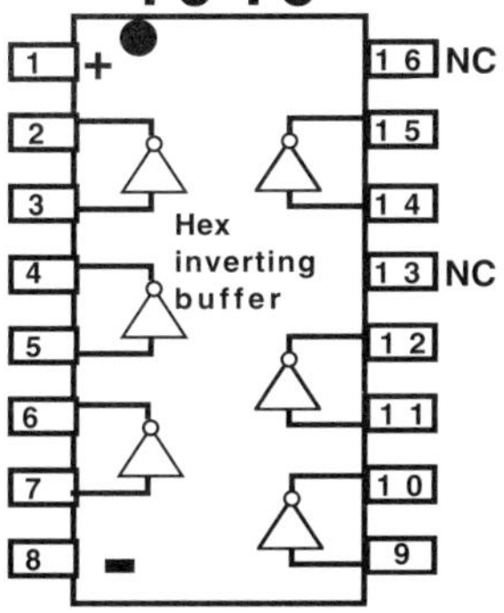

4050

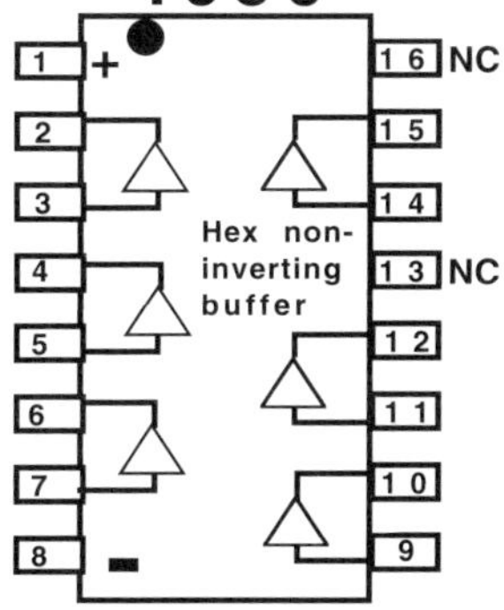

4069

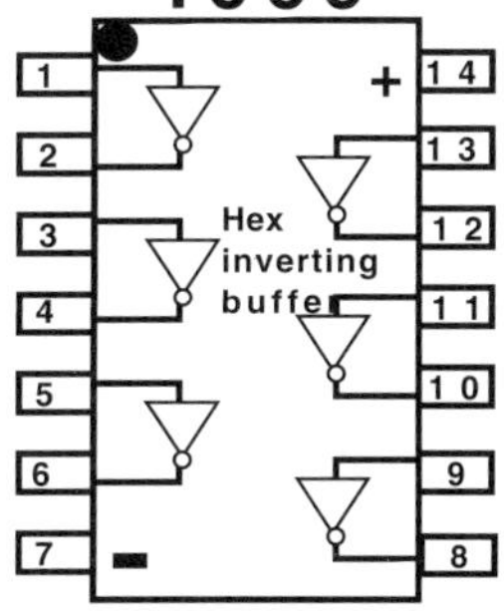

4502

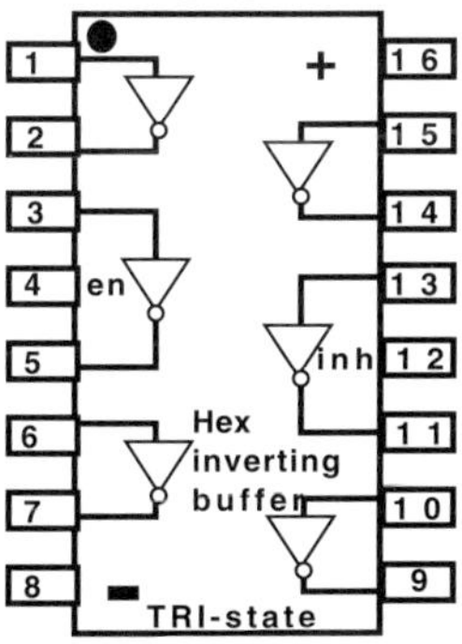

All six inverters can be used separately but they share a common enable (en) and inhibit (inh). If enable is 0 and inhibit is 0, they all function as normal inverters, i.e., the output has the state opposite that of the input. If inhibit is 1, all outputs will be 0. If enable is 1, all outputs will be open (high impedance state), i.e., the third logic state – tri-state.

4503

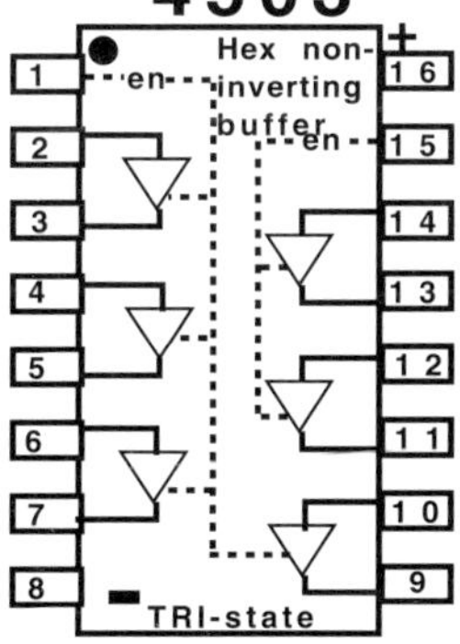

Here are six buffer gates that are non-inverting, i.e., the same logic state is present at both inputs and outputs. They are connected in one group of two and one of four. When enable is 0, the function is as above. When enable is 0, the output is open, i.e., the third logic state. Thus, this sub-circuit does not effect other logic in the circuit but is conceived as"transparent".

40106

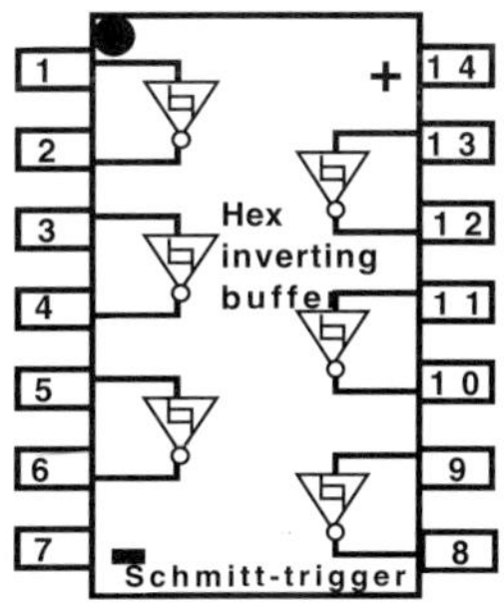

The **Schmitt-trigger** is a special variant of a buffer gate used when one needs a well-defined transition point between two logic states, e.g., from low to high or vice versa.
A normal buffer gate has an undefined transition area.

Case: Flat Pack for powering DC motors and stepping motors

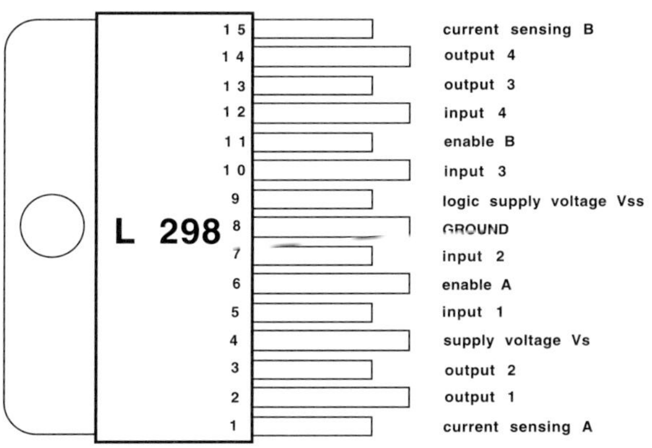

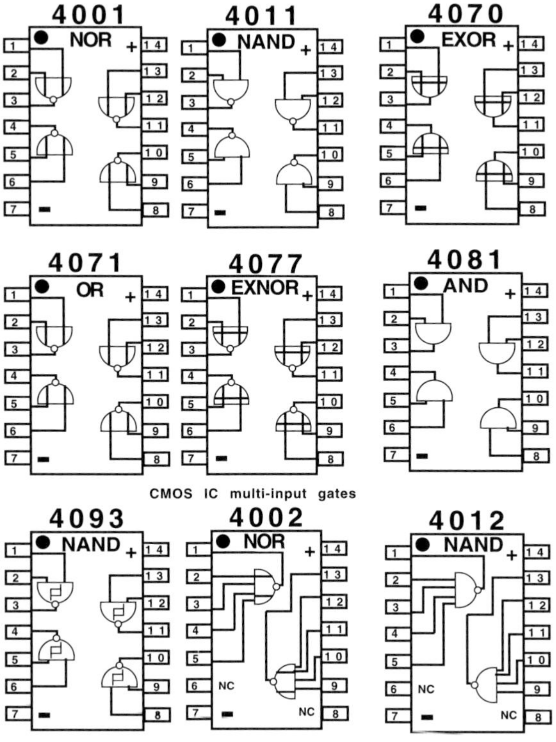

CMOS IC multi-input gates

Delta system summary:

Advantages:

-Delta is the starting point for a multi-train control system. You can run up to five trains on the same track block with just two wires to the layout.

-Delta is directly compatible with Märklin DIGITAL, when the layout and interest grow.

-Delta's power pack is nearly three times as powerful (30VA) as comparable starter set-ups.

-Delta engines have good low-speed characteristics:
- -start softly
- -change directions without jerks
- -change directions silently
- -change of direction without flashes

-Analog operation directly from the power pack for combining traffic with other conventional Märklin locomotives (without decoders) but on electrically insulated sections of the layout.

-Digital operations with addition of the train selector which then provides four-loco control.

A few disadvantages

-Analog engines cannot be run on a Delta layout – they would run at top speed only.

-Accessories such as turnouts, signals, etc. cannot be regulated by the Delta system. Regulation is accomplished with the usual tangle of cables as required by analog operations.

-Locos from Märklin's Digital series cannot use such extra functions as lighting (lights remain off) on Delta layouts.

Simply fill in name & address and mail it:

The response tag — your chance to get updates in the future

This book is meant to be a reference book, a manual and a book full of ideas for your Model Railroad Hobby. Our experience is that we will receive a lot of more advice, project improvements and corrections some time after the book has been released. Occasionally we even get totally new ideas on future projects to build. Maybe that is exactly what you are looking for.

This is absolutely no commitment on behalf of the author nor the publisher, but hopefully we will try to send you such information. All we ask of you is to fill in the tag below and mail/fax it to:

Allt om Hobby
P.O. Box 90 133
S-120 21 Stockholm
SWEDEN

telephone no: +46 8 99 93 33
fax no: +46 8 99 88 66
e-mail: freddy.stenbom@hobby.se (orders)
rutger.friberg@hobby.se (author)

I have Model Railroad Electronics 4. Please send me any update information when available to this address:

Name: ______________________________

Address: ______________________________

ZIP: ______________ State: ______________

Please send me the following books in this series:

—— Model Railroad Electronics 1
—— Model Railroad Electronics 2
—— Model Railroad Electronics 3

Don't hesitate to send us a copy of this page